Fear of Physics
And how to help students overcome it

Online at: https://doi.org/10.1088/978-0-7503-4866-9

IOP Series in Physics Education

The IOP Series in Physics Education aims to provide comprehensive, authoritative and innovative coverage for those that teach physics and related subjects at universities and other higher and further education institutions, and for those involved in physics education research.

Series Editor
Professor Peter Main
King's College London, UK

About the Editor
Peter Main obtained his PhD from the University of Manchester and, after post-docs in Manchester and Helsinki, he joined the University of Nottingham as a Lecturer in Physics in 1979. Following promotions to Reader and Professor, he eventually became Head of the School of Physics and Astronomy. His principal research interests were in quantum fluids and quantum transport in semiconductor and metallic heterostructures. He was also involved in many teaching innovations.

In 2002 he left Nottingham to join the Institute of Physics as Director of Education and Science. In this post he had overall responsibility for the Institute's work in education at all age levels, research, and diversity. Among many projects, he worked closely with Ofqual and awarding bodies on curriculum matters and with government to increase the number of physics teachers. He also initiated several projects improving the diversity of participation in physics.

In 2015 he joined King's College to become Head of Physics; he retains his interest in many projects in physics education and diversity.

About the Series
The IOP Series in Physics Education aims to provide comprehensive, authoritative, and innovative coverage for those that teach physics and related subjects at universities and other higher and further education institutions, and for those involved in physics education research.

The series supports evidence-informed professional practice and will cover topics including the following: assessment methods; feedback; conceptual understanding; problem solving; teaching methods; education technology; pedagogical theory; curriculum design; student engagement; misconceptions; employability; and social aspects of education.

Authors are encouraged to take advantage of electronic publication through the use of colour, animations, video, data files, and interactive elements, all of which offer particular benefits in communicating pedagogy.

Do you have an idea for a book you'd like to explore?
We are currently commissioning for the series; if you are interested in writing or editing a book please contact Caroline Mitchell at caroline.mitchell@ioppublishing.org.

A full list of titles published in this series can be found here: https://iopscience.iop.org/bookListInfo/iop-series-in-physics-education

Fear of Physics
And how to help students overcome it

Jeffry V Mallow
Loyola University, Chicago, IL, USA

Helge Kastrup
University of Copenhagen, Denmark

IOP Publishing, Bristol, UK

© IOP Publishing Ltd 2023

All rights reserved. No part of this publication may be reproduced, stored in a retrieval system or transmitted in any form or by any means, electronic, mechanical, photocopying, recording or otherwise, without the prior permission of the publisher, or as expressly permitted by law or under terms agreed with the appropriate rights organization. Multiple copying is permitted in accordance with the terms of licences issued by the Copyright Licensing Agency, the Copyright Clearance Centre and other reproduction rights organizations.

Permission to make use of IOP Publishing content other than as set out above may be sought at permissions@ioppublishing.org.

Jeffry V Mallow and Helge Kastrup have asserted their right to be identified as the authors of this work in accordance with sections 77 and 78 of the Copyright, Designs and Patents Act 1988.

ISBN   978-0-7503-4866-9 (ebook)
ISBN   978-0-7503-4864-5 (print)
ISBN   978-0-7503-4867-6 (myPrint)
ISBN   978-0-7503-4865-2 (mobi)

DOI   10.1088/978-0-7503-4866-9

Version: 20230501

IOP ebooks

British Library Cataloguing-in-Publication Data: A catalogue record for this book is available from the British Library.

Published by IOP Publishing, wholly owned by The Institute of Physics, London

IOP Publishing, No.2 The Distillery, Glassfields, Avon Street, Bristol, BS2 0GR, UK

US Office: IOP Publishing, Inc., 190 North Independence Mall West, Suite 601, Philadelphia, PA 19106, USA

Contents

Preface		ix
Author biographies		x
1	**Why physics?**	**1-1**
	Reference	1-2
2	**Student attitudes: mad science**	**2-1**
2.1	Student attitudes questionnaire [7]	2-2
	2.1.1 Attitudes questionnaire ([7] with permission of Springer Nature)	2-2
2.2	Student interviews [8]	2-5
	2.2.1 Interview questions ([8] with permission of Springer Nature)	2-5
2.3	Teacher interviews (Reproduced from [10] Copyright Morgan & Claypool Publishers)	2-7
	References	2-8
3	**Student anxieties: scary science**	**3-1**
3.1	The Science Anxiety Clinic	3-2
	3.1.1 Science anxiety questionnaire (Reproduced with permission from [9])	3-7
	References	3-15
4	**Constructivism: rational versus radical**	**4-1**
	References	4-7
5	**Relations and correlations**	**5-1**
	References	5-3
6	**Gender: not your grandmother's inequity**	**6-1**
6.1	Statistics	6-1
	6.1.1 US	6-1
	6.1.2 UK	6-2
6.2	Biological differences?	6-3
6.3	'It ain't what you don't know that gets you into trouble. It's what you know for sure that just ain't so.' (Mark Twain)	6-3
6.4	Proportions of females studying science at universities	6-4

6.5	Seminal research studies	6-5
	6.5.1 Tobias	6-5
	6.5.2 Seymour and Hewitt	6-5
	6.5.3 Hake	6-6
6.6	Raising the percentage	6-6
6.7	Where do we go from here?	6-8
	References	6-8

7 Nationality: does it matter?   7-1

7.1	PISA 2015	7-1
	7.1.1 Key features of PISA 2015	7-2
	7.1.2 Gender differences	7-4
	7.1.3 Attitudes	7-5
7.2	The seven levels of proficiency	7-5
	References	7-7

8 Math anxiety: the pipeline choker   8-1

8.1	Mathematics: the filter	8-1
	8.1.1 Mathematics and gender	8-1
	8.1.2 Changes	8-2
8.2	Similarities between math anxiety and science anxiety	8-2
8.3	Differences between math anxiety and science anxiety	8-4
	References	8-7

9 Laboratory pedagogy: you can't get it from a cookbook   9-1

9.1	The Archimedean law of buoyancy, version 1	9-2
9.2	The Archimedean law of buoyancy, version 2	9-4
9.3	The Archimedean law of buoyancy, version 3	9-5
9.4	Labs and IT	9-6
	References	9-6

10 Group project pedagogy: more heads than one   10-1

10.1	Some characteristics of group project work	10-1
10.2	Examples of projects: Denmark upper secondary schools	10-2
	10.2.1 Small group projects	10-2
	10.2.2 Large group projects	10-4

10.3	University group projects	10-5
	10.3.1 LUC projects	10-5
	10.3.2 RUC projects	10-7
10.4	Pedagogical theory	10-11
	References	10-12

11 Pedagogies for different student populations: they're not dumb, they're different   11-1

11.1	Evaluations	11-2
	11.1.1 Teacher–student classroom evaluation	11-2
	11.1.2 Teacher–student group project evaluation	11-4
	11.1.3 Teacher–teacher evaluation	11-4
	11.1.4 Student–teacher evaluation	11-5
	11.1.5 Student–student evaluation	11-5
11.2	Teaching physics students	11-5
11.3	Teaching pre-health students	11-6
11.4	Teaching liberal arts students	11-7
	11.4.1 Scientific methodology	11-8
	11.4.2 Newtonian physics	11-9
	11.4.3 Relativity	11-9
	11.4.4 Quantitative analytical skills	11-9
	References	11-11

12 Physics and society: close encounters   12-1

12.1	Close encounters of the first kind: the legendary cold inhuman physics lab	12-1
12.2	Close encounters of the second kind: physics as the confluence of the legendary nerds	12-3
12.3	Close encounters of the third kind: physics for society	12-4
	12.3.1 Provision of correct information	12-4
	12.3.2 Interaction with communities	12-4
	12.3.3 Research addressing societal issues	12-5
	12.3.4 Individual initiatives	12-5
12.4	Conclusion	12-6
	References	12-6

13 Information technology: use, under-use, over-use, misuse   13-1

13.1 Use   13-1
13.2 Under-use   13-5
13.3 Over-use   13-5
13.4 Misuse   13-5
    References   13-6

14 Online education: interaction at a distance   14-1

14.1 Methods of online education   14-2
    14.1.1 Videotaped teaching   14-2
    14.1.2 Teaching a few students with interaction   14-3
    14.1.3 Teaching a distant class   14-3
    14.1.4 Teaching students, each sitting at home   14-3
    14.1.5 Webinars   14-4
14.2 The online environment   14-4
    14.2.1 Bricks and mortar   14-4
    14.2.2 Social interaction   14-5
14.3 What can (and cannot) be done?   14-5
    References   14-6

Summary   15-1

Appendix: Questionnaires   16-1

Preface

This book addresses barriers to learning physics, and how to help students overcome them. Preconceptions and misconceptions about the subject often produce anxiety. For many years, we have studied the correlation between attitudes and anxieties about science in various student populations and across national and gender lines.

Fear of physics is in large measure a subset of general science anxiety. The first chapters thus focus on science anxiety, with various special observations about physics. Later chapters focus specifically on physics. The topics we address are:

- Students' attitudes towards science.
- Science anxiety.
- Constructivism philosophy, and pedagogy. What role it plays in shaping student attitudes. Radical constructivist attacks on science in general and physics in particular.
- Correlations of the above.
- Gender. Many women's distinct views of science and of how it is practiced, with considerable focus on physics.
- Nationality. International test results. Similarities and differences between American and Danish students' anxiety and attitudes.
- Mathematics anxiety. Correlation of fear of physics with fear of the accompanying mathematics. Mathematics anxiety as a barrier to the study of physics.
- Laboratory pedagogy. Experimental work as a source of anxiety for some students. Laboratory experiments designed to build confidence.
- Group project pedagogy. Project work as a single course, or as a part of other courses. Projects designed to build student confidence.
- Physics teaching for different student populations, including physics students, other science students, pre-health professions students, humanities and social science students, and education students (future teachers).
- Physics and society. Focussing on 'real world' problems to attract students who might otherwise avoid physics.
- Information technology. IT as an essential tool for physics education and research and a source of anxiety.
- Online education: 'Interaction at a distance.' Challenges of distance learning. Effect on teacher–student interactions, students' engagement in group work, communication by use of whiteboards, tablets, and other tools. The role of anxiety.

The book's pedagogical recommendations are based on research carried out by us and by others, seasoned by more than 50 years of physics teaching each. While covering some of the same topics as our earlier book, *Student Attitudes, Student Anxieties, and How to Address Them: A Handbook for Science Teachers*, the book focusses on specific components that comprise fear of physics, and specific recommendations for conquering it.

Author biographies

Jeffry V Mallow

Jeffry V Mallow is Professor of Physics Emeritus, Loyola University Chicago. His primary research area has been quantum mechanics. He discovered and named the phenomenon 'science anxiety'. Co-founder of the first university clinic to help students overcome this anxiety, he has published widely in the field that he initiated. Much of his work has been in collaboration with Helge Kastrup and other colleagues in Denmark.

Helge Kastrup

Helge Kastrup is Emeritus Professor of Mathematics and Physics at Copenhagen University College, and is currently a lecturer in science and mathematics at the University of Copenhagen. He has written research articles and books, including *Student Attitudes, Student Anxieties, and How to Address Them: A Handbook for Science Teachers* (2016) with Jeffry Mallow. Professor Kastrup is a contributor to several textbooks on the natural sciences.

IOP Publishing

Fear of Physics
And how to help students overcome it
Jeffry V Mallow and Helge Kastrup

Chapter 1

Why physics?

Over the last few decades, various teaching strategies have been shown to aid science learning. These include:

- The implementation of teacher–student and student–student interaction in lieu of straight lecturing.
- The design of laboratories to elicit creativity in approaches to problem definition and solution, in lieu of 'cookbook' instructions.
- The demonstration of multiple approaches to acquiring scientific knowledge, such as posing questions that may have more than one correct answer[1], or teaching several ways to present results.
- The recommendation of alternative presentations of materials from other textbooks, journals, and popular expositions.

But pedagogy is not enough. Students bring to the science classroom preconceived attitudes, as well as the emotional baggage called 'science anxiety'. Each of these or both can impede learning, even with the best teachers. There turn out to be connections between science attitudes and science anxiety. For example, students who regard science as inherently cold, unfriendly, and negative toward the individual, or think that science is inherently hostile and biased against women are likely to be science-anxious.

Much of this book elucidates what we have learned as science teachers and science education researchers. We have studied various groups in the US and Denmark. These included university students majoring in the sciences, mathematics, humanities, social sciences, business, nursing, and education; high school students; teachers' seminary students; science teachers at all levels from middle school through

---

[1] 'What it the specific resistance of copper?' permits one correct answer. 'How could you measure the electrical resistance of a length of copper wire?' allows for several.

university; and science administrators. We have also examined data and recommendations from the UK.

But why physics in particular? The easy answer is that we are physics teachers. This is what we know and that is from where we draw our various examples. However, there is more to it than that. 'Physics avoidance' is common, for example, in US high schools, often aided and abetted by teachers and administrators. Unlike biology and chemistry, it is not often required. Danish female gymnasium guidance counselors have sometimes invoked radical feminism—'physics is an oppressive Eurocentric male social construction'—to counsel female students that is their civil right to refuse to take physics courses. Of US university students, only one third of one percent are studying physics as their major subject. Even among the STEM (Science Technology, Engineering, and Mathematics) students, only two percent are majoring in physics [1]. In comic—and sometimes not-so-comic—stereotypes, mad physicists play a prominent role[2].

In our various questionnaires we have noticed that questions related to physics play a substantial role in shaping student attitudes and concomitantly provoking their anxieties.

By sharing both our research and our own experience over more than a century[3], we hope to help our fellow physics teachers understand students' attitudes about physics, to recognize the connections between these and their anxieties, and to see how pedagogies that take both into account can greatly assist in our students' learning.

Reference

[1] Nicholson S and Mulvey P J 2020 Physics bachelor's degrees: 2018 results from the 2018 survey of enrollments and degrees *AIP* https://aip.org/statistics/reports/physics-bachelors-degrees-2018

---

[2] Although to be fair, Victor Frankenstein was a biologist.
[3] Well, combined.

Chapter 2

Student attitudes: mad science

Attitudes are shaped in several ways. For science, popular opinion arguably tops the list, with books, films, and other media the proud culprits. Among classic mad scientist movies [1] we find any number of physicists. Here is a smattering.

The Invisible Ray: 'A scientist becomes murderous after discovering, and being exposed to the radiation of, a powerful new element called Radium X.'

Man-Made Monster: 'A mad scientist turns a man into an electrically-controlled monster to do his bidding.'

The Fly: 'A scientist has a horrific accident when he tries to use his newly invented teleportation device.'

Monster of Terror: 'A young man visits his fiancée's estate to discover that her wheelchair-bound scientist father has discovered a meteorite that emits mutating radiation rays that have turned the plants in his greenhouse to giants.'[1]

Last, but surely not least, Stanley Kubrick's *Dr Strangelove*. ('Mein Führer, I can valk!') Even the good guys, such as Spiderman, the Hulk, and at least a dozen more, gained their powers by exposure to radioactivity.

Wikipedia lists fictional scientists in books, films, TV, and video games. In each category, mad scientists are prominent [2]. Almost a third of a thousand horror movies shown in the UK from 1930 to 1980 featured mad scientists.

In C S Lewis's *Space Trilogy* [3], a soulless physics professor is nothing short of the incarnation of evil. (To which one may quote Austrian writer Robert Musil, trained as an engineer, in *The Man without Qualities*: 'A soul: what's that? It is easy to define negatively: it is simply that which sneaks off at the mention of algebraic series.' [4])

---

[1] Although it is not related to physics, who can resist *The Devil Bat*? 'A mad scientist (played by Bela Lugosi) develops an aftershave lotion that causes his gigantic bats to kill anyone who wears it.'

Live Science, an audience-supported online journal, lists among its 'Top Ten Mad Scientists' [5], Albert Einstein, Leonardo da Vinci, Nikola Tesla, J Robert Oppenheimer, Richard Feynman, Oliver Heaviside, and Freeman Dyson.

In the abstract to a 1988 article in *Physics Today* titled 'The physicist as mad scientist' [6], physicist Spencer Weart summarized the paradigm: 'Unstable scientists plotting to master and destroy can be found almost anywhere one looks in children's television and comics—and in a surprising amount of adult fiction as well .... To many people, then, the words "nuclear physicist" bring to mind a weird and evil picture.' Thus are attitudes shaped, aided and abetted by society at large, and thus is generated the baggage that students bring to the study of science.

2.1 Student attitudes questionnaire [7]

To examine what some of these attitudes are, we developed a student attitudes questionnaire. We administered it together with an anxiety questionnaire, which we will discuss in the following chapter. Approximately 1000 American and 900 Danish students filled out both. Below and in appendix A is a copy of the attitudes questionnaire. The majority of items either refer explicitly to physics or subsume it.

The respondents comprised groups from Loyola University Chicago: physics students, humanities students, algebra-based 'College Physics' students (many of them with pre-health career goals), education students; Danish students from three gymnasia[2], from Roskilde University (RUC), and from University College Copenhagen (UCC), a teacher training college[3].

2.1.1 Attitudes questionnaire ([7] with permission of Springer Nature)

Instructions. Please circle the number that best describes the degree with which you agree or disagree with each item below, using the following scale:

Strongly disagree	Disagree	Neutral	Agree	Strongly agree
1	2	3	4	5

1. Science reflects the social and political values, philosophical assumptions, and intellectual norms of the culture in which it is practiced.
2. Science is a 'level playing-field' in which men and women have equal status and opportunity.
3. Tomorrow's truths in science will be different from those of today.
4. It is possible for two scientists to carefully perform the same experiment and get very different results, each of which is correct.
5. Science has nothing to do with my life.

---

[2] The Danish gymnasium is about the same as the last two years of US high school and the first year of college. The curricula of the Danish gymnasia and universities are a fair match in content and level to those of the American groups listed.
[3] A small cohort of Chicago Public School science teachers also filled out the questionnaires. We shall limit our discussion to students.

6. Scientists agree on fundamental subjects like global warming, disposal of nuclear waste, and the use of stem cells.
7. Science is by its nature hostile to women.
8. Newton's laws of motion may eventually be modified.
9. Scientists' ideas apply to some physical objects in the Universe but not others.
10. The difference in number of men and women scientists is primarily due to biological differences.
11. The choice of topics for scientific research is affected by the views of the culture in which scientists work.
12. There are no such things as objective facts.
13. Science is boring.
14. The difference in number of men and women scientists is primarily due to differences in opportunity.
15. Science is inherently cold and unfriendly.
16. Science is a conspiracy between governments and scientific agencies formed to keep ordinary people from taking part in the democratic process.
17. Although interpretations can be ambiguous in things like personal relationships or poetry, in science the facts speak for themselves.
18. Newer scientific theories build on their predecessors.
19. Scientific experiments do not really discover 'the laws of nature', but instead let scientists find whatever they expect or want to find.
20. Women have a harder time succeeding in science than men do.
21. Modern scientists are responsible for most of the dangers in our world.
22. Science is a mental representation constructed by the individual.
23. When it comes to controversial topics such as which foods cause cancer, there's no way for scientists to evaluate which scientific studies are the best: everything's just a matter of opinion.
24. Every scientific theory is eventually proved completely wrong, and must be discarded.
25. The scientific view of the world is just an agreement among scientists.
26. Despite what scientists would have us believe, science is actually subjective.
27. Science transcends national, political, and cultural boundaries.
28. Scientists don't understand normal people.
29. The natural world can best be explained through a combination of perspectives, including science, paranormal phenomena, and astrological horoscopes.
30. The difference in number of men and women scientists is primarily due to individual choice.
31. The scientific knowledge in use today may be obsolete tomorrow.
32. Scientific laws work well in some situations but not in others.
33. Current ideas about particles that make up the atom will always be maintained as they are.
34. Objective facts are an illusion.

35. I cannot fulfill my need for creativity within the closed framework of the natural sciences.
36. Science is a naturally male field.
37. Scientific theories are simply agreements among scientists.
38. Current ideas about particles that make up the atom apply to physical objects everywhere in the Universe.
39. The reason fewer females than males choose careers in science is that women don't want to be restricted to the narrow scientific way of understanding the world.
40. The results of scientific research experiments are affected by the views of the culture in which scientists work.

Many of the questions seem repetitive. This standard approach of questionnaires is to assess consistency. The questionnaire was designed to probe a variety of attitudes. They can be broken down into several categories. The claim of item 33, that current ideas about particles that make up the atom will always be maintained as they are, might be called 'very traditional'. So is the claim in item 2, that science is a level playing-field, in which men and women have equal status and opportunity. A more correct term for these claims is 'demonstrably false', as all scientists know, or should.

More moderate and more accurate are such items as 27. Science does transcend national, political, and cultural boundaries. Correct results do not depend on who is doing the study, although, as per item 11, the choice of topics for scientific research is affected by the views of the culture in which scientists work. Those choices are often determined by what governments think are important. Scientists would generally agree on this item.

At the other extreme are attitudes such as those exemplified by items 7, 15, 19, and 34, known as 'radical constructivism'. (In chapter 4, we shall discuss constructivism both as a theory and a pedagogy.) They fall into three broad categories:

- Subjective construction of knowledge; i.e. that all scientific knowledge is 'socially constructed' and therefore fundamentally (not simply in how it might be misused) a suspect mode of inquiry.
- Negativity of science toward the individual; i.e. the belief that science and scientists are in some sense fundamentally 'anti-human'.
- Inherent bias against women, i.e. that science is fundamentally gendered (not simply in how it may be practiced discriminatorily), that there is a difference between 'female science' and 'male science', and that science as currently practiced is male science, oppressive to women.

We found that these three categories transcend culture. The Danish students described science in the same ways as the American students.

So what generalizations can we make? First, judging by the scholarly and popular literature, the three categories of radical constructivism are articulated by citizens of numerous nations. Large numbers of people around the world believe them. Second, gender plays a major role. In almost every category, the responses of females in both countries differed significantly from those of males. Third, radical constructivist views about science are linked to science anxiety in both the US and Denmark. We may expect therefore that females would demonstrate more science anxiety than males, and we would be correct. We will discuss this further in chapter 3 and subsequent chapters.

Physics is typically viewed as the least comprehensible of the sciences and the scariest, with its applications the most malevolent, as per Weart's example above. It is the science subject least required in secondary schools or universities. Its practitioners are often described by item 28 above. (Who of us has not had the experience of being asked 'So what you do?', and as soon as we say the Ph word, the response is something like, 'I could never understand that' or, 'How can I even talk to you?') Those of us in the US who remember the 1970s may recall anti-Vietnam-War teach-ins versus anti-nuke rock concerts.

2.2 Student interviews [8]

In addition to the attitude questionnaire, we carried out interviews with a subset of the respondents, a cohort of 55 students: 25 Americans and 30 Danes. The full list of questions appears below and in appendix A.

2.2.1 Interview questions ([8] with permission of Springer Nature)

1. What choices for advanced study will you make, and why?
2. How many of your teachers in science have been male? Female? How has that affected you? Do they teach differently?
3. Do you find any of the sciences dry? If so, which ones and why?
4. Has your mathematics experience had an effect on your attitudes towards science?
5. What causes of anxiety in science can you identify?
6. Why do you think gender differences exist in subject choices?
7. How do your non-science friends view you? OR How do you view science students?
8. How are you affected by others' views of scientists and science students? Is this gender related?
9. What types of gender interactions have you experienced in science classes?
10. What do you think of single-sex educational groupings?
11. Do you experience anxiety in any subjects or situations?
12. Some people have suggested that there are no such things as objective facts and that science is simply constructed from the personal opinions and subjective beliefs of scientists. How do you feel about this particular viewpoint?
13. Some people have suggested that science is inherently hostile and biased against women. How do you feel about this particular viewpoint?

14. Some people have suggested that science is inherently cold, unfriendly, and negative toward the individual. How do you feel about this particular viewpoint?
15. Some people argue that science is one way of describing the natural world among others to be considered as well; for instance, astrology, creationism/intelligent design. What is your opinion of this?
16. Some people argue that all of the science of today will become obsolete and forgotten in centuries to come. What is your opinion of this?

The items were covered in random order, chosen by the interviewees. They cover a range of topics, including attitudes towards science, scientists, and science students. Here we consider the ones regarding attitudes towards the nature of science itself: 12–16. (That being said, students chose to include remarks therein about the practice and practitioners of science, such as those related to gender, which we will consider in chapter 6.)

For the combined Danish–American cohort, question 15 had the highest frequency of responses, followed in order by questions 13, 14, 12, and 16. We shall discuss question 13 further in chapter 6. We cannot say for certain, but based on our observations and experiences, it seems to us that radical constructivism has taken less hold in Denmark than in the US. Science students in both national groups rejected the pseudoscientific views expressed in question 15. Danes were somewhat less likely to believe in the hostilities described by questions 13 and 14.

While belief in creationism and intelligent design were not prevalent among our interviewees, it was not entirely absent. Most worrisome was its articulation by some American primary education students, our future teachers. Unlike constructivism, this is a debate that obviously transcends academic circles. The notion that simply presenting science will automatically discredit pseudoscience will not do.

The preponderance of responses to arguments for radical constructivism (*grosso modo*, 'there are no facts') were negative and independent of national group, gender, and course of study. While there were several caveats, these constituted examples of 'moderate' constructivism as generally accepted by scientists—theories are not immutable, but *are* data based.

All respondents of both genders in the Danish and American groups answered 'No' to the question of inherent hostility to women. Numerous respondents pointed out the cultural bias of males and male-dominated science to females, and the societal stereotypes which feed such bias (male 'mad scientists' for example). Some thought that it was due not to the nature of science, but rather to females' lack of interest in science. In general, respondents in both national groups claimed (accurately) that this was changing, with more females entering the sciences. That being said, the proportions of US female students earning bachelor's degrees in physics in 2018 was 21%, far lower than in chemistry (close to 50%) and in biology (over 60%) [9]. We shall discuss this further in chapter 6.

2.3 Teacher interviews (Reproduced from [10] Copyright Morgan & Claypool Publishers)

Following the student interviews, we carried out interviews with five physicists. Three were Danish and two were American. Each was an expert in science education; each had been both a teacher and an administrator. They ranged in age from 50s to 60s. The three Danes have been physics and mathematics teachers in gymnasia. Two have also had university teaching experience. Each has been a physics advisor in the Danish Ministry of Education, and two of the three have held administrative positions in education beyond physics. The two Americans have been department chairs and deans; one was also director of an institute. We shall label them respectively DK1, DK2, DK3, US1, and US2. Since the questions we posed dealt with the full range of topics in the student interviews, and the responses did not separate them, we defer the full discussion to chapter 5. Here we give a summary of observations about the students' perceptions of the nature of science, interview questions 12–16.

DK1:

- Most Danish students had not been introduced to such provocative ideas.
- Non-science students' claims that science is cold and unfriendly is still due to the way it is taught.
- Giving credence to pseudoscience is unknown among the science students, but some non-science students are open to the notion.

US1:

- Students found the issue of science's intrinsic hostility to women easy to discuss, because they didn't believe it; in fact, they thought it was silly.
- As regards science's hostility towards everyone, this view oscillates over time as responses to, for example, lunar landings versus nuclear accidents and antibiotic failures.
- The question of pseudoscience had bifurcated the US student population, and that accounted for its popularity as a topic of discussion. Intelligent design should be treated in science classes. The obligation of the teachers is to debunk it on scientific grounds.
- Views of astrology versus astronomy are not generally considered controversial. It may well be that many people think they are the same.
- By and large, the notions addressed in questions 12, 14, and 15 were just wrong and not worthy of discussion.

US3: There has been little change over the years in the prevalence of pseudoscience. He described his experience as a graduate student in Texas, where he received a threatening phone call because of his public opposition to anti-science.

Several of the educators elected to discuss question 16 in particular. They saw this misconception as an opportunity to teach their students about the evolving nature of science. 'How science has changed', 'the dynamic nature of science', 'building on prior knowledge' were the expressions they used. DK1 argued that being interested in this question is a prerequisite for being a good science teacher.

Of all of the themes discussed by the educators, gender played the most prominent role. We shall discuss this further in chapters 5 and 6.

References

[1] lordvanxar 2016 All classic mad scientist movies *IMDb* https://imdb.com/list/ls032247233/
[2] https://en.wikipedia.org/wiki/
[3] Lewis C S 1938 *Out of the Silent Planet* (London: The Bodley Head)
 Lewis C S 1943 *Perelandra* (London: The Bodley Head)
 Lewis C S 1945 *That Hideous Strength* (London: The Bodley Head)
[4] Musil R 1930–43 *The Man Without Qualities* (Austria: Rowohlt Verlag)
[5] Live Science Staff 2008 The top 10 mad scientists *Live Science* https://livescience.com/11380-top-10-mad-scientists.html
[6] Weart S 1988 The physicist as mad scientist *Phys. Today* **41** 28
[7] Bryant F, Kastrup H, Udo M, Hislop N, Shefner R and Mallow J 2013 Science anxiety, science attitudes, and constructivism: a binational study *J. Sci. Edu. Technol.* **22** 432–8
[8] Mallow J, Kastrup H, Bryant F, Hislop N, Schefner R and Udo M 2010 Science anxiety, science attitudes, and gender: interviews from a binational study *J. Sci. Edu. Technol.* **19** 356–69
[9] American Physical Society Bachelor's degrees earned by women, by major *APS* https://aps.org/programs/education/statistics/womenmajors.cfm
[10] Kastrup H and Mallow J 2016 *Student Attitudes, Student Anxieties and How to Address Them: A Handbook For Science Teachers* (San Rafael, CA/Bristol: Morgan & Claypool/Institute of Physics)

Chapter 3

Student anxieties: scary science

The psychology of anxiety was considered as early as 1844 by the Danish philosopher Søren Kirkegaard in his book *Begrebet Angest* (*The Concept of Anxiety*) [1]. He distinguished between fear (frygt) and anxiety (angest). Fear deals with things of which there is good reason to be afraid, such as a viper in your bed. Anxiety on the other hand means being scared of something that is not intrinsically fearful. Science anxiety is not a viper in your bed. But to a science-anxious person, it feels like it. Many students suffer from science anxiety, affecting their studies, excluding them from careers where science and math are prerequisites, and leaving them uninformed citizens whose political decisions are not based on good information or careful analysis.

In 1977 one of us (JVM) recognized the phenomenon for which he coined the term 'science anxiety'. It often manifests itself as panic on examinations in science classes, but it is distinct from general test or performance anxiety. Science anxiety and the techniques for its alleviation were developed by him, working with psychologists, into the Loyola Science Anxiety Clinic, the first of its kind [2].

The causes of science anxiety are many, including past bad experiences in science classes, science-anxious teachers in elementary and secondary schools, a lack of role models, gender and racial stereotyping [3–6], and the stereotyping of scientists in the popular media.

Science anxiety and science enrollments are affected by role models or the lack thereof. For example, science anxiety and its gender differences begin in primary school. Two studies examining science anxiety in students showed that as soon as science was formally introduced into the curriculum, when students were about nine years old, the girls exhibited science anxiety to a greater degree than the boys [7, 8].

So is science anxiety an inborn deficiency that the pupil has to accept as narrowing his or her potential course of study? No. There are ways to diminish rather than create or enhance anxiety. Here we will describe what we discovered in the clinic about students' science anxiety and how to deal with it. Then we will describe what we found in large-scale studies in our respective countries, Denmark and the US. Throughout we discuss what steps physics teachers can take to help their students gain confidence.

3.1 The Science Anxiety Clinic

Here is a list of the anxiety-producing statements articulated by students in the clinic:

1. No matter how much I study, I'll never understand science.
2. If I don't get a perfect score, it means I didn't study enough.
3. If I'm not studying all the time, I'm not studying enough.
4. I should have studied more.
5. I probably studied all the wrong material.
6. I'm not worried enough about this science exam: I'll probably do poorly.
7. If this problem looks easy, it's probably because I don't really understand what is asked for.
8. I'll never get this lab experiment to work.
9. I could get an A in this science course if only I worked hard enough.
10. There's just too much material to study here. I'm overwhelmed.
11. I should know this material.
12. If I can't understand this science material, I shouldn't even be in school.
13. All this trouble I have grasping science just proves how incompetent I am.
14. If I don't do well in science, I'm worthless.
15. If I ask the teacher a question, I'll show how dumb I am.
16. What will the teacher think of me if I can't answer this question?
17. Other students ask such intelligent questions. Why can't I?
18. Other people seem to know the secret of understanding science. Why don't I?
19. Everyone around me is writing answers to this exam. I can't think of a thing.
20. Everyone else understands the material.
21. I'm not really as smart as the other students.
22. Everyone else can do it except me.
23. Science is too hard for normal people.
24. Only super-brains can understand science.
25. Boys are supposed to be good in science.
26. Science is not for girls.
27. Science is not for me.
28. I'm scared of science, and I'm all alone in my fear.
29. I can't be expected to know this.
30. I don't have a scientific mind.
31. I don't really have to know any science to be successful in my chosen career.
32. If I miss this problem on the exam, I may fail it. Then what if I also fail the next exam? And the next?

33. If I don't get an A in science, I won't get into medical school (dental school, nursing school, graduate school, college, etc).
34. My future hinges on this science course.
35. My folks are counting on me to do well.
36. I should be studying, but I want to play.

Our technique for addressing these concerns consisted of categorizing the statements, analyzing their content, then providing 'coping statements' to help the students overcome their anxieties. The teacher may use this method in one-on-one situations, such as during office hours. But first the teacher must ascertain whether the student is in fact science-anxious. This may not be so easy. The student may come in with a different 'presenting agenda'; for example, 'I just don't do well on tests'. Here the teacher may ask if this is true for all courses (general performance anxiety: 'stage fright'), or just for the science course. If the latter, then the teacher can probe further about the negative self-statements. These techniques have been taught to science teachers in workshops. WARNING! We are not psychologists. If the student seems to have other problems, the teacher must stop IMMEDIATELY. He or she is free to recommend counseling at the student health center or psychological clinic. The student may refuse, or even be offended, but the teacher will have done their job.

But what if this is not science anxiety, but rather (or in addition) math anxiety? They are not the same, but they may overlap. One way to ascertain this is to ask if the student has anxiety in math classes as well as science classes. We discuss this in detail in chapter 8.

Let us see what characteristics these 36 statements exhibit that make them negative and anxiety producing. First of all, they are filled with loaded words, words that take science learning and turn it into an emotionally trying event. Words such as 'should', 'never', 'expected', 'worthless', 'incompetent', and 'perfect' occur throughout. These words distract students from focusing on the task of comprehending science, because they take them out of the here-and-now and put them into some hypothetical threatening future situation, such as an exam or some other place where they will be shown up for their incompetence.

Second, the statements introduce a measure of irrationality into an otherwise rational situation; namely, studying science. The irrationality is introduced in any number of ways. Statements 1–10 might be described as examples of abdicating one's sense of judgment. Instead of recognizing the rational aspect of studying science—that it takes work, but is not magical and that the students can decide when they have studied enough—they ascribe all the power to some vague external force that will never allow them to grasp science. They see themselves as powerless—powerless to decide when to stop studying, powerless to make an experiment work, powerless to know what to study, powerless to decide when a question is easy or hard. So they revert to a primitive mode of trying to affect the outcome of an event perceived to be out of their control; namely, praying to an idol. The modern form of this is called worrying, as in statement 6. Not the ordinary, reasonable sort of worrying, but a kind of institutionalized worrying, as if enough worry will improve their chances of success. Statement 10 represents the science-anxious person who simply gives up. Rather than focusing on science study as a series of separate tasks taken one at a time, the anxious person sees only the forest, not the trees; that is, only the whole mass of material, which of course appears overwhelming.

Statements 11–16 are irrational in a different way. They hook the students' performance in science directly to their sense of personal worth. Each difficult task is taken as a sign of basic worthlessness. Interestingly, among science-anxious students each success in a science course is ascribed to luck. (This seems to be especially true for females.) So it is easy to see how irrational these statements are. Just as statements 1–10 are examples of abdicating the power to make rational judgments, statements 11–16 are examples of abdicating the sense of self, tying it entirely to performance.

Statements 17–28 might be titled 'making comparisons with a mythical norm'. In these statements they compare themselves with others, doing so on the basis of little information or misinformation. In statements 17–22, they see others as normal, able to understand science, relaxed, and intelligent, while they are abnormal. These kinds of negative self-statements are highly irrational, because the science-anxious person actually has little evidence that everyone else is smarter. In fact, the word 'everyone', like 'never', is a tipoff that the statement is irrational. How, for example, does one decide, as in statement 17, that other students' questions are intelligent? Either you do not understand the question, in which case you cannot assess whether it is intelligent, or you do understand the question, in which case the questioner is not any smarter than you! In fact, statement 17 is really saying what statement 7 said: 'if I understand it, it can't be very profound, and if I don't understand it, then it is probably profound'. Statements 23–27 are the mirror images of 17–22. In these latter statements, the people who can understand science are the abnormal ones, while the student is in the normal group, the group that cannot understand science. Statements 25 and 26 are of this sort when expressed by females. When expressed by males, they are in the same category as statements 17–22. In either case, the students define themselves as intrinsically in the wrong group, with no hope of passing into the other. Of course, in statements 23–27 there is the consolation that they are not alone and are 'normal'. But it's not much consolation.

Statement 28, the last of the 'mythical comparison' statements, is a very popular one with science-anxious people. It completely and utterly isolates them from the rest of the world. Only they have a fear of science, and no one else can understand their problem. As one might imagine, this worldview produces enormous anxiety. (One great virtue of the Science Anxiety Clinic was that it quickly undercut this irrational statement, because the room was filled with science-anxious people.) Statements 29–36 might be categorized as 'adding to the pressure'. The first three are the most deceptive because they look like statements to relieve pressure, to minimize the importance of learning science. If they were made in an emotionally neutral setting, then they might be accurate, albeit unfortunate, assessments of the situation. For example, an art student looking at an advanced biology textbook could accurately say, 'I can't be expected to know this' or 'I don't need to know science for my career'. Even the statement 'I don't have a scientific mind', although misguided, need not produce anxiety when uttered by this hypothetical art student. However, these statements are usually voiced in a very different context. The student who is required to take a science course or the adult who encounters something technical that is job related—these are the people who seek refuge in those statements. But there is no refuge to be found. They are expected to know the material, it does affect their career, and the claim that they don't have a 'scientific mind' is irrelevant.

Statements 32–36 lay the pressure on in a much more direct way. The first three are prime examples of what psychologists call 'catastrophizing': extrapolating from a limited situation, such as a science exam, to an irrational and terrifying set of consequences. The panic evoked by this sort of negative self-statement tends to paralyze the person at just the time when efficient performance in science is essential. The last two statements are less connected to any particular science event, such as an exam. However, they tend always to be there in the background when the student is trying to accomplish any science task—reading a textbook, working problems, or just studying science material. They therefore provide a continual level of distraction, which makes it harder to concentrate.

Of course, this is not an exhaustive list of negative self-statements. Nevertheless, the common types of statements have been well represented. We should also note that not all the statements appear to be science related and, in fact, some students generate these statements about other subjects. We are not talking just about math, which can generate anxieties similar to those about science, but about, for example, foreign languages. However, the prevalence and intensity of these statements appear to be much greater in people facing the prospect of having to learn science.

Once students have identified their negative self-statements, they must begin the search for the antidote, the so-called 'coping statement'. This is not simply the opposite of the negative statement. It does no good, for example, to tell the person who is scared of science (statement 28) that 'there's nothing to be afraid of'. This simply discounts the fear and replaces it with the idea that the anxious person is a little crazy. Other sorts of positive-sounding statements, such as 'you can do it if you try' or 'don't worry about grades, just do your best' may be well meaning, but they are useless. They are, in fact, merely a form of cheerleading that does not address the fear itself.

The basic requirement of the coping statement is that it focus on what is rational and what is irrational about the negative statement, and separates the one from the other. Let us take negative statement 1 as an example: 'No matter how much I study, I'll never understand science.' What is rational? That it takes serious study to understand science. What is irrational? The mystical prediction of the future embodied in the words 'no matter how much' and 'never'. A good way to approach an appropriate coping statement is, therefore, to question the irrational assumptions: '"Never" is an absolute and final word. How can you predict the future? On what evidence do you base the assumption that no matter what you do it will prove fruitless? How can you be so certain of failure?' Thus we see that the pessimism embodied in the negative self-statement is not countered by some unrealistic optimist but rather by an objective observer who can focus on the irrational beliefs underlying the negative self-statement and attack these beliefs with reason.

We must emphasize at this point: it is not sufficient to 'think through' science anxiety. People do not necessarily change the way they feel and act because they understand something intellectually. Hence, even after students have found coping statements (no small task in itself), it takes time for these new messages to integrate themselves into their mental frame of reference and replace the negative statements.

Coping statements, like negative statements, vary from person to person, so it would not be particularly useful to run down our list of negative statements and provide a coping statement for each one. In fact, the most useful thing for the science-anxious person to do is to try to develop personal coping statements. We can, however, provide some general guidelines. The first has already been stated: find the rational core and the irrational trimmings of the negative statement. Then counter the irrational part with objective statements, rather than with cheerleading statements. The next thing to do is to change the semantics; namely, the way students tend to phrase things. Look for statements involving 'should' and 'must' as well as generalizations such as 'never' or 'always' or 'everyone but me'. The student needs to eliminate them actively from their vocabulary—they are not rational. They must begin focusing on the here-and-now. A good objective coping statement useful for almost any negative statement is, 'This negative statement is distracting me from the task at hand. It does not help in learning science. Focus on the task.'

For the first ten negative statements, the ones we call 'abdicating your sense of judgment', coping statements naturally take the direction of recapturing this sense. 'No amount of studying guarantees perfection; some amount of studying is certainly enough; worrying guarantees nothing at all; if I have studied, then some problems will look easy.' These are the sorts of directions that coping statements can take to help students recapture their sense of judgment. Sometimes a textbook or lecture seems confusing because it is confusing—poorly written, poorly presented, or even wrong. Students should entertain this possibility as they attempt to study science, in the same way as they do when they study other subjects.

Statements 11–16 must be countered by coping statements that separate self-worth from performance: 'What does it mean to be "worthless"? Isn't this an irrational word? Isn't the teacher there to answer my questions? Isn't asking questions a good way to learn?' And suppose the teacher is unresponsive, or even, as unfortunately happens from time to time, sarcastic? Whose problem is that? The students' or the teacher's? Students have done nothing wrong by trying to understand science. Help must be sought elsewhere, such as from another teacher.

Statements 17–28, the comparisons with the mythical norm, are best countered by questioning the basis on which they are made: 'What evidence do I have that the other person's questions are clever and mine are not? Isn't it rather magical to think that everyone but me knows the secret of understanding science? Who says that only certain people—super-brains, or boys—are the only ones who can understand science? What is the evidence to support this claim?' (There is none.)

The catastrophizing statements can be dealt with by fantasizing the worst possible consequences of, say, doing poorly on an exam and allowing the fantasies to become as grandiose as possible—and then recognizing how irrational they are: 'One event does not shape my future forever after. What is the most awful thing that will happen if I can't work this physics problem? Is that realistic? How terrible are the actual consequences?' These sorts of questions make up the kinds of coping statements that counteract the catastrophic negative self-statements that produce science anxiety in highly charged situations such as exams.

It is sometimes the case that once students have been able to counter and neutralize some negative self-statements, others begin to surface. This is quite natural: getting in touch with the things we tell ourselves is not a one-shot deal. It is more like peeling an onion, with successive layers appearing after we have dealt with the outer ones. So if one set of negative statements gives way to another, students need to keep using the same techniques, developing new coping statements. They need to be patient. Nothing happens overnight. But the techniques outlined here have one great advantage: they work. And they have worked on many science-anxious people. They will very likely work for any given student. To believe otherwise is itself a negative self-statement.

Are there objective measurements of science anxiety? Yes. Here is our questionnaire. (A copy is in appendix A.)

3.1.1 Science anxiety questionnaire (Reproduced with permission from [9])

The items in the questionnaire refer to things and experiences that may cause fear or apprehension. After each item, place a number that describes how much YOU ARE FRIGHTENED BY IT NOWADAYS.

0	1	2	3	4
Not at all	A little	A fair amount	Much	Very much

1. Learning how to convert Celsius to Fahrenheit degrees as you travel in Canada.
2. In a philosophy discussion group, reading a chapter on the categorical imperative and being asked to answer questions.
3. Asking a question in a science class.
4. Converting kilometers to miles.
5. Studying for a midterm exam in chemistry, physics, or biology.
6. Planning a well-balanced diet.
7. Converting American dollars to English pounds as you travel in the British Isles.
8. Cooling down a hot tub of water to an appropriate temperature for a bath.
9. Planning the electrical circuit or pathway for a simple 'light bulb' experiment.
10. Replacing a bulb on a movie projector.
11. Focusing the lens on your camera.
12. Changing the eyepiece on a microscope.

13. Using a thermometer in order to record the boiling point of a heating solution.
14. You want to vote on an upcoming referendum on student activities fees, and you are reading about it so that you might make an informed choice.
15. Having a fellow student watch you perform an experiment in the lab.
16. Visiting a science museum and being asked to explain atomic energy to a twelve-year-old.
17. Studying for a final exam in English, history, or philosophy.
18. Mixing the proper amount of baking soda and water to put on a bee sting.
19. Igniting a Coleman stove in preparation for cooking outdoors.
20. Tuning your guitar to a piano or some other musical instrument.
21. Filling your bicycle tires with the right amount of air.
22. Memorizing a chart of historical dates.
23. In a physics discussion group, reading a chapter on quantum systems and being asked to answer some questions.
24. Having a fellow student listen to you read in a foreign language.
25. Reading signs on buildings in a foreign country.
26. Memorizing the names of elements in the periodic table[1].
27. Having your music teacher listen to you as you play an instrument.
28. Reading the theater page of *Time* magazine and having one of your friends ask your opinion on what you have read.
29. Adding minute quantities of acid to a base solution in order to neutralize it.
30. Precisely inflating a balloon to be used as apparatus in a physics experiment.
31. Lighting a Bunsen burner in the preparation of an experiment.
32. A vote is coming up on the issue of nuclear power plants, and you are reading background material in order to decide how to vote.
33. Using a tuning fork in an acoustical experiment.
34. Mixing boiling water and ice to get water at 70 °F.
35. Studying for a midterm in a history course.
36. Having your professor watch you perform an experiment in the lab.
37. Having a teaching assistant watch you perform an experiment in the lab.
38. Focusing a microscope.
39. Using a meat thermometer for the first time, and checking the temperature periodically until the meat reaches the desired 'doneness'.
40. Having a teaching assistant watch you draw in art class.
41. Reading the science page of *Time* magazine and having one of your friends ask your opinion on what you have read.
42. Studying for a final exam in chemistry, physics, or biology.
43. Being asked to explain the artistic quality of pop art to a seventh grader on a visit to an art museum.
44. Asking a question in an English literature class.

---

[1] For example, recognizing Cu as copper. Not all of them, as in the Tom Lehrer song https://www.youtube.com/watch?v=AcS3NOQnsQM.

Developed in the Science Anxiety Clinic, the questionnaire has subsequently been utilized in numerous controlled studies of large groups of students, both American and Danish [10]. We analyzed it in two ways. The first is by the standard multiple regression analysis. We found three correlations with science anxiety. They are non-science anxiety, then gender, and finally, choice of course of study. See figure 3.1. There were no significant correlations with nationality.

The correlation with course of study is especially interesting, for two reasons. The first is: did the non-science students choose their course of study for positive reasons or because they wanted to avoid science? We have some evidence that the latter is the case, based on short informal questionnaires. The second is, were the males telling the truth? If science is still perceived as a masculine field, then the non-science male students may have been ashamed to admit that they chose their course of study to avoid science. For example, in the Danish gymnasium until recently, students had to select one of two 'tracks': math or language. The former included the sciences; the latter, all of the humanities and social sciences. About the same number of females as males chose the math track. But females overwhelm males by about two to one in the gymnasium. So the language track consisted of a large majority of females and just a few males. In one group we studied, there were only six males—and they were at pains to tell us that they weren't gay!

A second method for measuring science anxiety, in fact what we may designate as acute science anxiety, is what we call the 'bucket analysis'. Instead of using the full 1–5 scale, we divided the answers into two buckets. Levels 1, 2, and 3 we took as not science-anxious, or minimally so. Levels 4 and 5 we designated as acute science anxiety. At first we simply looked for gender differences in science anxiety for the

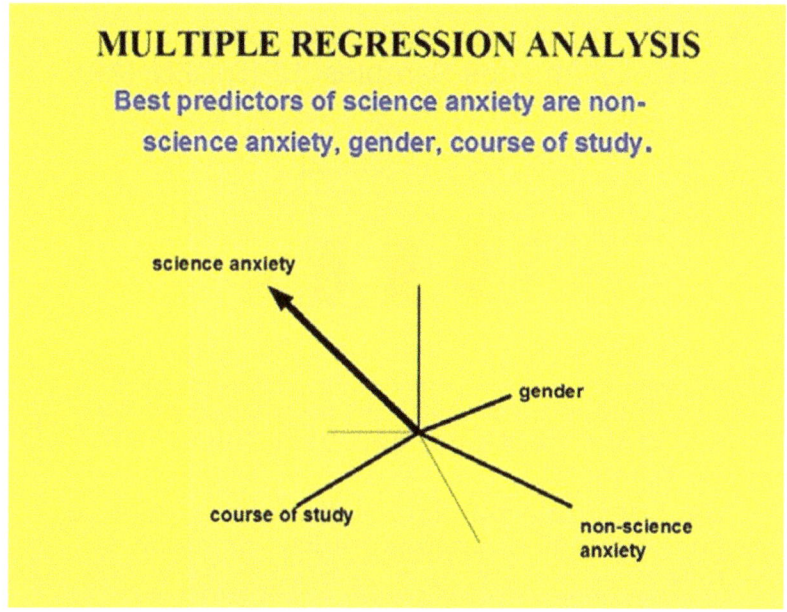

Figure 3.1. The results of a multiple regression analysis of a large-scale science anxiety survey.

whole student cohort. We didn't find any. We then thought about the multiple regression results and realized that the best predictor of science anxiety—let us call it SA—was general (science and/or non-science) anxiety—we will call this GA. What this meant is that the large group of non-anxious students were masking any effects. So we separated out that group and looked only at the generally anxious group to see how strongly they exhibited science anxiety. We found significant gender differences, with females showing more anxiety than males. Then, using as a measure the ratio of acute science anxiety to acute general anxiety, i.e. SA/GA, and performing a chi-square analysis, we found that 32% of generally anxious Danish males exhibited acute science anxiety, compared to 71% of Danish females, 77% of US males, and a whopping 90% of US females. See figure 3.2.

But what does it mean to have acute science anxiety? How many 4s and 5s are necessary? We chose the number to be one. This on the face of it seems artificially low. But would two be better? Five? Ten? We don't think so. First of all, a student with even a single item showing acute science anxiety is worthy of anxiety reduction intervention. Second, we also looked at how many of those who chose one, chose more. The numbers for the various groups ranged from about two-thirds to over ninety percent. So we are confident that our choice represents for the most part more than one manifestation of the anxiety.

Now we can ask, in what courses of study are the students most science-anxious? The results for US students are shown in figure 3.3. The order from highest to lowest was: business, education, humanities, mathematics, nursing, science, social science, and undeclared. In all cases, the proportion of females was higher than that of males. There are some features here that merit further discussion. One is that even science

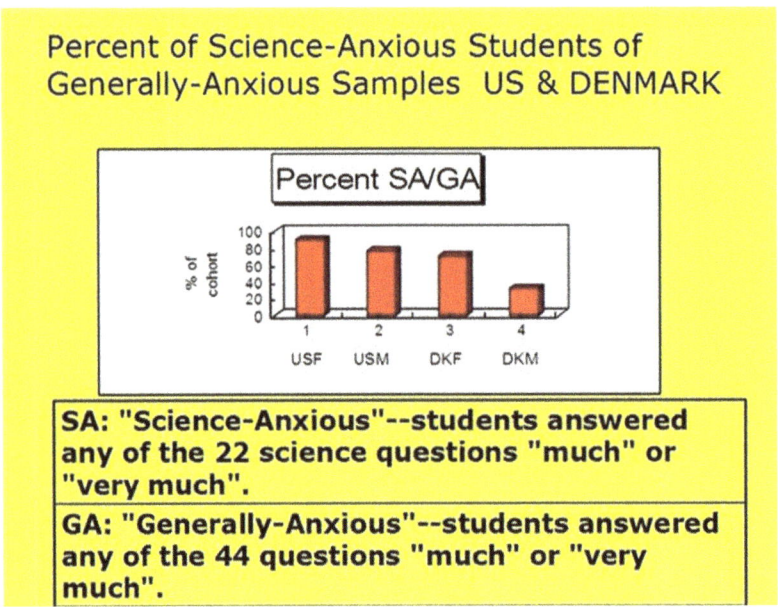

Figure 3.2. Percent of science-anxious students of generally anxious samples, results for US and Denmark.

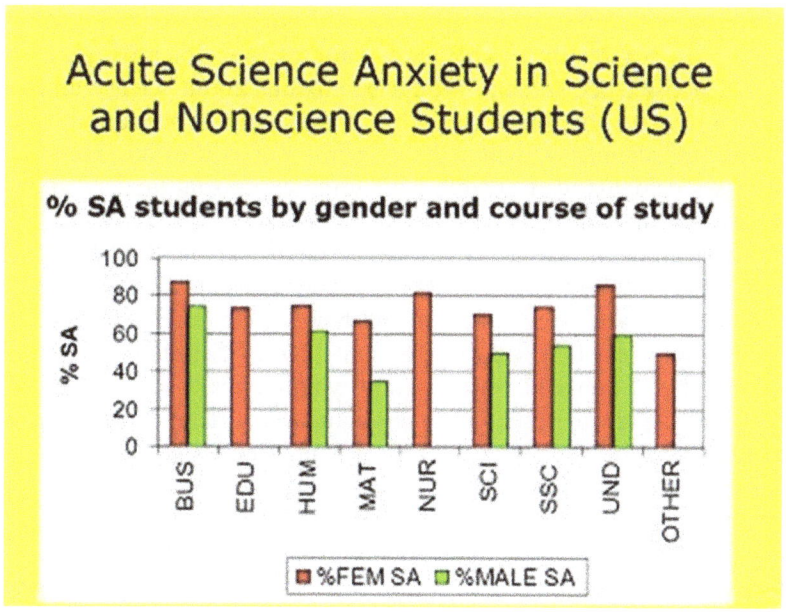

Figure 3.3. Acute science anxiety in science and non-science students (US). The ordinate is the ratio of acute science anxiety to acute general anxiety: SA/GA.

majors suffer from science anxiety. In fact, many of the students in the original Loyola Science Anxiety Clinic were science majors. It might be biology students who feared physics, or even a few physics students who loved physics but were afraid of it[2]. A second feature of note was that math majors were found near the top of the anxiety scale. It is unlikely that so many of them also suffered from math anxiety, confirming that the two anxieties were distinct.

A third feature, perhaps the most disturbing, is the high proportion of education majors, almost all female, suffering from science anxiety. It is they, our future teachers, who will be tasked with introducing young students to science. It is they who will pass on their own anxiety, often in the form of avoidance[3]. We noted in chapter 2 that they also tended more than other cohorts to believe in pseudoscience.

In chapter 2 we discussed the interview studies with students. There we restricted our discussion to attitudes. Here we address the responses to the following questions on anxiety:
- 5. What causes of anxiety in science can you identify?
- 11. Do you experience anxiety in any subjects or situations?

---

[2] This was in fact the case for JVM, who barely got through his college physics courses due to panic. His teachers just thought he wasn't smart enough. He went on to earn a doctorate in astrophysics. Some years later, as an assistant professor, he founded the first science anxiety clinic so that students would not have to go through what he did.

[3] When JVM enrolled his son in kindergarten, he asked the teachers if they taught any science. The answer was yes: on Friday afternoon. Furthermore, they admitted that only one of them taught it, and it was biology.

Both Danish and American groups discussed both questions. The US students stayed very close to relating experiences, not causes. Interestingly, the simple equation that non-science students would be anxious about science and vice versa did not turn out to be the case. Physics students expressed anxiety about mathematics and chemistry, while the responses of education, humanities, and College Physics students were mixed, with some expressing anxiety about science and some about non-science.

Some Danish students claimed that it was primarily teachers who were the cause of science anxiety. In addition, Danish non-science students attested to their fear of 'one correct answer', and science students to the opposite: fear of subjective answers in non-science courses. Or as JVM's Senior Physics Seminar instructor, Nobel Laureate Polykarp Kusch said, 'I don't know what a physics seminar really is. In an English seminar the teacher gives an opinion and the student gives an opinion, and then they discuss it. But in physics, damn it, you're either right or you're wrong.' Some students questioned this distinction, with examples of exact humanities such as foreign languages, and inexact science, such as open questions about genetic technology and nuclear power. The reasons for the national differences are not clear, but must have something to do with the educational systems or societal norms overall.

We should not read too much into the details, not least because exact translation from one language to another is impossible. But one thing stands out: the influence of teachers. It is they who most affect the anxiety (or lack thereof) of the students.

We then assessed American and Danish teachers' classroom behaviors. The questionnaire used for this was based on the one shown in appendix A. It was part of a workshop designed by a committee of the American Association of Physics Teachers (AAPT). Titled *Developing Student Confidence in Physics* [11], it was designed to show teachers how they could reduce anxiety and raise confidence among their students. The teachers were asked to assess their relative focus on content versus relationship. The desired outcome of the workshop was to have them strive for balance between the two, to cover the material while maintaining a classroom atmosphere that minimized anxiety and maximized confidence. Originally administered to the workshop attendees, the questionnaire was later extended to other AAPT members. It was translated into Danish, shortened and modified, and administered to attendees of national meetings of the Danish Physics Teachers' Association. This version, which appears below, was then administered to attendees of AAPT conferences. It comprises a selection of five of the twelve scenarios in the AAPT version, plus two new questions, numbers 6 and 7, added by the Danish teachers.

Binational teachers' personal self-inventory[4]
The questionnaire is designed to help you assess your teaching style, and consider modifying it if you deem it necessary.

---

[4] The American version appears in appendix A.

Circle the letter or letters of the alternative(s) that most nearly describe your behavior in each situation. Please circle at least one response for each situation.

1. The lab performance of your students seems to be getting worse. More of them are not completing the experiments during the lab period. You would:
   A. Emphasize the importance of laboratory work and the necessity for accomplishing the lab assignments.
   B. Use conversation to show your concern and to encourage the students.
   C. Discuss the function of experimental work in science and work with the students to improve the quality of their performance.
   D. Let the students be responsible for themselves.

2. Your students bombed the first hour-long exam. You have stressed the grading standards for the course. Classroom participation is beginning to improve. Now the second exam is coming up. You would:
   A. Engage in helpful friendly interactions and continue to make sure all students are aware of your standards.
   B. Be satisfied with the improvement and take no further action.
   C. Praise the class for its improved participation and let each student know you care about his/her work.
   D. Emphasize the importance of good study habits and the completion of each homework assignment.

3. You are teaching the course for the first time. The students are doing badly on the exams; perhaps the course is moving too fast. You have decided to make some changes. You would:
   A. Invite class discussion in developing the change. Do not direct.
   B. Slow down and see what happens.
   C. Allow class to formulate the basis for change in classroom activities.
   D. Incorporate group recommendations but direct change.

4. Students have been complaining that the subject is too abstract. Some of them have asked that you include some applications in the course. You would:
   A. Pick out applications for inclusion in your lectures.
   B. Ask them to come up with a list of applications of interest to them and from this list they select the three most relevant. You then include these topics in your lectures.
   C. Ask the class for some applications of interest to them. You pick out a few for inclusion in your lectures.
   D. Explain to them that science is based on abstract reasoning and that once they understand the basic principles, the applications will be easier to understand.

5. Many of the students concentrate poorly in your Friday afternoon class. You would:
   A. Accept poorer concentration in Friday afternoon classes.
   B. Ask the students for suggestions on ways to increase class concentration, and use those suggestions that are consistent with course goals to stimulate interest.

C. Inform students you will include material on exams which is covered during Friday classes.
   D. Contact individual students and encourage them to be more active in the Friday classes.
6. You are preparing a lecture for tomorrow on some scientific law. You remember that this topic has proved difficult for students in previous years, and you are determined to improve your approach. You would:
   A. Employ a more rigorous proof of the law, and tell the students that examples similar to those done in class would appear on the exam.
   B. Omit the proof of the law, devote the entire period to solving examples, and say nothing about how the topic will be covered on the exam.
   C. Prepare questions about the law to ask the students in class. Praise students who answer correctly and tell students who don't that most students have difficulty with this topic.
   D. Do a rigorous proof very slowly, taking pains to describe every step. Then ask for volunteers to explain assigned examples during a later class, and assure the volunteers that you are available to help them prepare their presentations.
7. You are grading a set of homework problems, on which almost all of the students have made the same mistakes. You would:
   A. Ask the students to discuss in class why their approach seemed reasonable to them and gently reminds them of your rules on collaboration on homework.
   B. Solve the problem correctly in class the next day.
   C. Correct all the mistakes on the papers and ask the students to pay careful attention to your written comments when you return the papers the next day.
   D. Ask the students to prepare corrected versions of their solutions and discuss these corrected versions with each student individually.

Scoring

Teachers' selections of the scenarios were divided in to those emphasizing relationship and those emphasizing content. There are no 'right' or 'wrong' answers. The goal is to strive for balance.

Item	Relationship		Content	
1	C	B	C	A
2	A	C	A	D
3	C	A	C	B
4	C	B	C	A
5	B	D	B	C
6	C	D	A	D
7	A	D	C	D

We found no obvious differences in the teaching methods of the two populations [12]. But there was a difference in the American versus the Danish students' confidence levels, as we saw in figure 3.2. One explanation may be that the constant exposure to science from the early grades on renders Danish students more confident than their American peers. Another possibility is that each class through primary school has the sane teacher. This long-term teacher–student relationship may well produce confidence in any subject, including science.

Finally, we may ask if simply taking a physics course could change a student's level of science anxiety, regardless of the student's major subject. The answer is yes (and happily in the right direction). A study [13] of several Loyola student cohorts taking various introductory physics courses for a semester showed a reduction in science anxiety, especially acute science anxiety. As was the case in the Science Anxiety Clinic, non-science anxiety was reduced as well.

References

[1] Kierkegaard S 1844 *Begrebet Angst* (self-published)
[2] Mallow J V 1986 *Science Anxiety: Fear of Science and How to Overcome It* (Clearwater, FL: H & H Publishing)
[3] Mallow J V and Hake R 2002 Gender issues in physics/science education (GIPSE)—some annotated references, 300 references and 200 hot-linked URLs http://physics.indiana.edu/~hake (GIPSE-3a.pdf)
[4] Hake R and Mallow J 2008 Gender issues in science/math education (GISME). Over 700 annotated references and 1000 URLs: part 1—all references in alphabetical order, online as an 8.6 MB pdf at http://bit.ly/gXdrvR; and part 2—some references in subject order, online as a 4.9 MB pdf at http://bit.ly/fBTzqV. Available on request
[5] Bachelor's degrees in physics and STEM earned by women *APS Physics* http://aps.org/programs/education/statistics/womenstem.cfm
[6] Katemari R, Blue J, Hyater-Adams S, Cochran G L and Prescod-Weinstein C 2021 Resource letter RP-1: race and physics *Am. J. Phys.* **89** 751–68
[7] Chiarelott L and Czerniak C 1985 Science anxiety among elementary school students: an equity issue *J. Educ. Equity Leadersh.* **5** 291–308
[8] Chiarelott L and Czerniak C 1987 Science anxiety: implications for science curriculum and teaching *Clear. House* **6** 202–5
[9] Alvaro R 1980 The effectiveness of a science therapy program upon science-anxious undergraduates *PhD Dissertation* Loyola University Chicago, IL, USA
[10] Mallow J V 1994 Gender-related science anxiety: a first binational study *J. Sci. Educ. Technol.* **3** 227–38
[11] Fuller R, Agruso S, Mallow J V, Nichols D, Sapp R, Strassenburg A and Allen G 1985 *Developing Student Confidence in Physics* (College Park, MD: AAPT) workshop manual
[12] Mallow J V 1995 Students' confidence and teachers' styles: a binational comparison *Am. J. Phys.* **63** 1007–11
[13] Udo M K, Ramsey G P, Reynolds-Alpert S and Mallow J V 2001 Does physics teaching affect gender-based science anxiety? *J. Sci. Educ. Technol.* **10** 237–47

Chapter 4

Constructivism: rational versus radical

Constructivism has come to mean different things to different people. What all would agree on is its claim that we construct from our experiences our own personal understanding of the world we live in. We construct our own mental models as a framework for categorizing, analyzing, and registering our primary sensory input. Learning is therefore our internal saving of information in existing models or modification of the models when necessary. This is hardly a controversial notion; nevertheless, various definitions of what is called constructivism can easily be confusing and quite abstruse. Let us consider some and appraise them. We shall focus on those aspects that affect science attitudes, both negatively and positively.

The fact that our knowledge and understanding of the world is formed from data acquired through the senses has led philosophers of all ages to articulate many different versions of what has been called 'idealism'. The most common form is: 'The world is a construction of our brain.' It stands in contrast to the materialist view, which forms the basis of science: 'The real world exists and we experience it through our senses.' One form of the modern social constructivist theory of learning/cognition/scientific knowledge is an example of idealism. It claims that science is entirely a social construction made by individuals in their social context [1].

Our attitudes questionnaire of chapter 2 spans the spectrum from traditional empiricism through moderate constructivism to radical constructivism. For traditional empiricism we have the example of item 33, that current ideas about particles which make up the atom will always be maintained as they are. We might call this radically traditional[1]. Moderate constructivism is exemplified by item 8: 'Newton's laws of motion may eventually be modified[2].' We designated as radical constructivist

---
[1] And wrong.
[2] And have been: relativity and quantum mechanics.

such items as 7: 'Science is by its nature hostile to women.', 15: 'Science is inherently cold and unfriendly.', 19: 'Scientific experiments do not really discover "the laws of nature", but instead let scientists find whatever they expect or want to find.', and 34: 'Objective facts are an illusion.' That is, that science is *inherently* sexist, inhuman, and subjective. They stem from the academic movement usually described as 'post-modernism' with its claim that all knowledge, including science, is a social construction. We are not referring to the sometimes discriminatory *practice* of science, such as using health data acquired from white European males to generalize to other populations. We refer to the claims that the results of scientific measurement depend on, *inter alia*, nationality, gender, ethnicity, or social status.

There are post-modernist claims in other fields, such as equal validity of contribution of author and reader in literature, sometimes caricatured as, 'Shakespeare could fail a course in Shakespeare.' But science, especially physics, is a particular target. Post-modernists would argue that this is not the case, that they are simply subjecting science to the same criteria as other fields. We think otherwise: science is the elephant in the room of the post-modernist project. Its claims to objective knowledge must therefore be undercut. That can only be accomplished by arguing that there is no such thing as knowledge that is not socially constructed. The project has been advanced by the inauguration of a field known as Science Studies[3]. There is no requirement that its practitioners know any science. They use mathematical and scientific symbols and equations, generally incorrectly, to prove (not just analogize) the connection between those fields and psychology, sociology, history, and other fields, usually focusing on racism, sexism and other oppressions putatively intrinsic to science. Physics is their favorite target, with mathematics close behind. While the contents of the attacks are often risible, we should be cognizant of their effect. The post-modernists in the academy have produced disciples, who have then produced more disciples, and so on[4], until the ideas have acquired cachet, whence students' acceptance of radical constructivist statements in our questionnaire.

Beginning in the 1990s, the attacks on science led to the Science Wars. Prominently representing the scientists were the authors of three books: physicist Alan Sokal and physicist and philosopher of science Jean Bricmont [2], philosopher of science Noretta Koertge [3], and biologist Paul Gross and mathematician Norman Levitt [4]. (Sokal is famous for what is now called the Sokal hoax [5]).

Sokal and Bricmont focus on French philosophers, providing close readings of their claims, and unmasking them. For example, Jacques Lacan [6] claims that $\sqrt{-1}$ is associated with the erectile penis. Luce Iriguray [7] hypothesizes that $E = mc^2$ is a 'sexed' equation because it 'privileges the speed of light over other speeds that are vitally necessary to us'. Iriguray also posits that male physicists have privileged the study of rigid body mechanics over fluid mechanics because the former is a symbol

---

[3] We refer here only to the post-modernist movement. Traditional science studies has long been a respectable field, peopled primarily by philosophers who have made major contributions to the understanding of how science works, and have taken the trouble to actually learn some.

[4] Dare we call it a chain reaction?

of the erection, while fluids are associated with the feminine[5]. (This popular postmodernist view is dissected by Paul Sullivan [8].) This may at first glance seem amusing, but it is a direct attack on physics. Feminist sociologist Sandra Harding is notorious for her statement about the *Principia*: '[W]hy is it not as illuminating and honest to refer to Newton's laws as "Newton's rape manual" as it is to call them "Newton's mechanics?"' [9]. (She later said she regretted using the term [10]. That falls well short of disavowing it.) Evelyn Fox Keller argues that male physicists' cognitive development as children affects whether or not they support the Copenhagen interpretation of quantum mechanics [11]. She also quotes psychological studies of male physical scientists that claim to show that they 'tend overwhelmingly to have been loners as children, to be low in social interests and skills, indeed to avoid interpersonal contact' [12]. They 'are not very interested in girls, date for the first time late in college, marry the first girl they date, and thereafter appear to show[6] a rather low level of heterosexual drive' [13]. In accepting the notion that these, if they actually exist, are intrinsic characteristics, Keller ironically ignores social influences from childhood on, *inter alia*, mocking (geek, mad scientist, 'Built any bombs lately?'), isolation ('Ooh, a scientist. I can't talk to you!'), outright prejudice ('Scientists have no soul'[7].), and the difficulty of finding marriage partners who are interested in who they are and what they do.

The argument that Nature is irrelevant, that what matters is only the social interaction between and among groups of people, is a fundamental belief of Science Studies adherents[8]. Leading advocates of the reigning doctrine are Stanley Aronowitz and Bruno Latour [14], who unequivocally state that science is nothing more than a manifestation of political arrangements with the capitalist ruling class, and has no intrinsic validity. Thus, we have students agreeing with our questionnaire item 19: 'Scientific experiments do not really discover 'the laws of nature', but instead let scientists find whatever they expect or want to find.' In other words, there are no such things as facts—except for the post-modernists' facts[9].

A particularly egregious example is provided by a TV-program in Denmark. The highly regarded professor of astrophysics, Anja Andersen, was asked to discuss the flat earth hypothesis with a journalist. Unknown to her, the journalist had a proselyte of the Flat Earth Society in his ear plug directing the show. Once she became aware of this, Anja tried to stop the show being broadcast, as she felt abused. She paraded a T-shirt in the news, with the text '*Viden er faneme ikke et*

---

[5] Presumably those who suffer from erectile dysfunction study amorphous materials.

[6] A purported research study with the words 'appear to show'? And how did they measure that?

[7] Austrian author Robert Musil (1880–1942), trained as an engineer, remarks in *A Man Without Qualities*, 'A soul. What's that? It is easy to define negatively: It is simply that which sneaks off at the mention of algebraic series.'

[8] Loyola's chair of sociology once announced at a dinner that all science is a social construction. To JVM's reply, 'It's 68 °F in this restaurant' came the rejoinder, 'Only if we agree to it'. Hearing the story, a physicist colleague proposed that we take the sociology chair to the roof of the ten-storey building in which his department is housed, and challenge him to demonstrate that gravity is a social construction. It turns out that Sokal uses a similar analogy in an article revealing his hoax—and he lives on the 21st floor.

[9] How does one resist the temptation to respond to a post-modernist's claim that there are no such things as facts with 'Is that a fact?'

synspunkt' ('Knowledge is dammit not a point of view'). She was quoted as saying 'I was so indignant that they put it as two points of view against each other. But knowledge is, dammit, not a point of view. It is knowledge. I was upset that facts were treated as a matter of opinion, and that facts and fake news as two equally valuable views.'

Demonstration of a complete lack of scientific knowledge is no barrier to the postmodernist claims. Among them are Paul Virilio's descriptions of acceleration and decelerations as positive and negative velocities [15]; the invocation by Latour of a third observer who receives the data from observers in two inertial reference frames[10] [16]; the invention of meaningless pseudoscientific terms such as Jean Baudrillard's 'hyperspace with multiple refractivity' [17]. (This kind of thing has sometimes been called 'physics envy' [18].)

Koertge [19] describes in detail some of the effects of post-modernism on science and mathematics curricula. For example [20], a course in a university department lists the following objectives:

1. Describe the political nature of mathematics and mathematics education.
2. Describe gender and race differences in mathematics and their sociological consequences.
3. Examine the factors influencing gender and race differences in mathematics.
4. Critically evaluate eurocentrism and androcentrism in mathematics.

One can imagine that a sociology course would have such content. *But this is a course in a mathematics department.*

Koertge gives another example [21], this falling under the rubric of 'female friendly science'. The call for gender equity in course materials originally meant, for example, giving balanced word problems; e.g. projectile motion should sometimes be about volleyball rather than football[11]. Newer textbooks show that this has been taken seriously. But post-modernists have gone further. A statistics course includes the following problem:

> In sixty-five percent of all rapes, the victim knows her assailant. If we interview twenty women who were raped, what is the probability that no more than four of them were raped by strangers?

In addition to the question of how this can possibly be construed as female friendly, it is pedagogically misguided. The goal of physics and math problems is to teach students how to eliminate the 'story' and draw out the essentials. This would do just the opposite.

Post-modernists claim that their work is misrepresented because it is so profound that scientists don't understand it. Scientists disagree [22].

---

[10] What frame is *this* observer in?
[11] Basketball has been gender-equitable for several decades.

The claim that science is nothing but a social construction is a misrepresentation of Thomas Kuhn's [23] model of scientific progress as going through stages where a specific paradigm is accepted. Kuhn's term 'paradigm shift' refers to the observation that such shifts are often not dependent on the quality of then-current measurement. Ptolemy's epicycle model[12] of the Solar System gave more accurate predictions of planetary positions than did Copernicus's heliocentric model. The new Copernican paradigm was later modified (Keplerian ellipses rather than circular orbits, and subsequent improvements) and fit the data better, but the original shift to the Copernican model was philosophical. A common statement of this misconception is, 'everything is relative'[13]. Were that the case, the validity of scientific measurements could well depend on, for example, gender, race, or ethnicity. Indeed, more than a few radical constructivists claim there *are* different forms of science, such as women's science and men's science, Eastern science and Western science, Black science and White science[14].

A related misstatement is 'every scientific theory is eventually proved completely wrong, and must be discarded' (questionnaire item 24). If this were true, then all of Newtonian physics would have had to be discarded in favor of Einstein's theories of relativity. This misuse of Kuhn ignores the fact that a new theory is only accepted insofar that it encompasses the old.

If it were indeed the case that science is simply a social construction, then scientists would be guilty of concocting theories, writing papers, and publishing books just as a way of making a living [24][15]. Yet this is a not uncommon notion, and it was reflected in some of the positive answers to our attitudes question 19.

What then is the form of constructivism that makes sense in science [25]? Philosophically, constructivism was presented as a counter to empiricism. Empiricism separates facts from theory, and claims that facts determine the validity of theory. A popular empiricist epigram is 'a theory is only as good as the last experiment that didn't contradict it'. A good example of empiricism is the measurement of the aberration of starlight during the 1919 solar eclipse, which validated the predictions of Einstein's general theory of relativity. It is one of the few examples of an important scientific discovery that made headlines worldwide. It made Einstein 'the

---

[12] In the general public, one often talks of *theories* where one should say *models*. The so-called Big Bang 'theory' is a very good model of the Universe. But it is continually being revised. Dark matter, dark energy, and inflation were not part of the Big Bang model in the early 1980s. The explanatory theory is Einstein's general relativity. Similarly, it was Newton's theory of gravitation that explained the heliocentric model although it could not account for the phases of Venus, as the heliocentric model could. Before Copernicus different models were used to explain the apparent luminosity of planets and their places in the sky.

[13] Einstein is reputed to have regretted his choice of nomenclature, wishing instead that he had called it the Theory of Invariance.

[14] Such categorization has a sordid provenance. Here is German Nobel Laureate Phillip Lenard in the earlier part of the last century: 'Jewish physics can best and most justly be characterized by recalling the activity of one who is probably its most prominent representative, the pureblooded Jew Albert Einstein. His relativity theory was to transform and dominate all physics; but when faced with reality, it no longer had a leg to stand on. Nor was it intended to be true. In contrast to the equally intractable and solicitous desire for truth of the Aryan scientist, the Jew lacks to a striking degree any comprehension of truth.'

[15] JVM has a T-shirt that proclaims 'I BECAME A PHYSICIST FOR THE MONEY AND THE FAME'.

science genius' overnight. It is also the prime example of a theory that would have been discarded or radically altered if Eddington's measurements had not been in accord with Einstein's predictions.

Constructivism, on the other hand, claims that the test of valid science is that it provides a general organization of theory and experiment to lend an overarching interpretation: coherent, consistent, and wide-ranging. In short, it generalizes to a philosophy; i.e. the simple statement we made at the beginning: we construct from our experiences our own personal understanding of the world we live in. The role of facts is to hone these interpretations and make them more discriminating. Thus, in many situations, a single factual observation cannot serve as a logically crucial experiment. For example, the question, 'Is light a wave or a particle?' turned out to be an inappropriate theoretical construction. Experiments confirmed both, forcing reconsideration of the question, and the replacement of classical Newtonian physics with quantum physics and its own newly constructed theoretical categories (with the caveat that the latter reduce to the former in its range of validity). The constructivist reorganizing of the body of physical knowledge into the theory of quantum physics is subject to empirical tests of its validity. Hensen *et al*'s [26] measurements of Bell's inequality demonstrated that Bohr's interpretation of quantum mechanics accorded with experimental data, while Einstein's did not.

Many of the items in our attitudes questionnaire address philosophical constructivism; namely, the perceived nature of science, as well as pedagogical constructivism; namely, the means by which scientific knowledge may be discovered and communicated. As we have discussed at length above, the former can be problematic. The latter takes as its basic premise the simple notion that learning is an active process. From this viewpoint, students must be taught to be active learners, to construct their own understanding, rather than be treated as empty vessels into which the instructor pours knowledge. The teacher's obligation is to bring students into situations where they have to construct new patterns of thinking for themselves.

Effective science teachers tend to be philosophical empiricists and pedagogical constructivists. They believe in facts, but also know that learners actively construct knowledge, in which facts play an important but not exclusive role. Constructivist science pedagogy has become widespread [27]. Its most common manifestation is what in the US is called 'interactive engagement' (IE), the substantial departure from lecture mode to student–teacher and student–student interaction. In Britain and Denmark, it is called 'inquiry based science teaching' (IBST) [28]. A seminal study by Hake [29] demonstrated that introductory physics courses with an IE emphasis led to significantly higher student learning than traditional 'chalk-and-talk' lectures. A less well-known but equally important study by Gautreau and Novemsky [30] tested IE experimentally in separate sections of a single physics course at one institution, and found the same result. Half of the instructors were told to teach by lecture, the other half by IE. The next semester, the same instructors reversed teaching methods. IE produced better results regardless of the teacher[16].

---

[16] This turned out to have a sad end. Even given the evidence, inertia prevailed. Instructors went back to lecturing.

A study of students' performance in physics courses [31] showed that IE improved understanding of physics concepts and reduced the gender gap. Traditionally, the field of physics education research [32] has based its pedagogical initiatives on constructivist assumptions, and classroom practices such as IE confirm the value of this approach.

There is one additional problem. Who is capable of knowledge construction, and at what level? Jean Piaget [33], often described as the founder of pedagogical constructivism, divided intellectual development into four stages: sensory–motor, ages 1–2; pre-operational, 3–10; concrete operational, 10–15; and formal operational, 15 and above. (These numbers are of course approximate.)

In Piaget's original presentation the development through the four stages was linear; i.e. it depended only on age. We now know that there is more than one variable. In fact, one can be in different stages at the same time if, for example, one is anxious. Science anxiety can cause a person who is normally formal operational to drop down to concrete operational. This might be observed as panic on exams. We now know that this is not generalized anxiety, but a particular manifestation of science anxiety. Thus, with factors such as science anxiety in play, it is difficult or even impossible to tell if new insight is really gained.

In the next chapter we will show how pedagogical constructivism, correctly applied, can modify attitudes and reduce anxiety.

References

[1] Kastrup H 2007 About constructivism *AGORA* **10** 21–9
[2] Sokal A and Bricmont J 1998 *Fashionable Nonsense: Postmodern Intellectuals' Abuse of Science* Intellectual Impostures (London: Profile)
[3] Koertge N (ed) 1998 *A House Built on Sand: Exposing Postmodernist Myths about Science* (Oxford: Oxford University Press)
[4] Gross P and Levitt N 1994 *Higher Superstition: The Academic Left and Its Quarrels with Science* (Baltimore, MD: Johns Hopkins University Press)
[5] Sokal affair *Wikipedia* https://en.wikipedia.org/wiki/Sokal_affair
[6] Sokal A and Bricmont J 1998 *Fashionable Nonsense: Postmodern Intellectuals' Abuse of Science* Intellectual Impostures (London: Profile) ch 2
[7] Sokal A and Bricmont J 1998 *Fashionable Nonsense: Postmodern Intellectuals' Abuse of Science* Intellectual Impostures (London: Profile) ch 3
[8] Koertge N (ed) 1998 *A House Built on Sand: Exposing Postmodernist Myths about Science* (Oxford: Oxford University Press) ch 5
[9] Harding S 1986 *The Science Question in Feminism* (Ithaca, NY: Cornell University Press)
[10] Nemeck S 1997 The furor over feminist science *Sci. Am.* **276** 99–100
[11] Fox Keller E 1995 *Reflections on Gender and Science* (New Haven, CT: Yale University Press) ch 7
[12] Roe A 1953 *The Making of a Scientist* (New York: Dodd, Mead)
Roe A 1956 *The Psychology of Occupations* (New York: Wiley)
[13] McClelland D C 1962 On the dynamics of creative physical scientists *The Ecology of Human Intelligence* ed L Hudson (Harmondsworth: Penguin)

quoted in ed E Fox Keller 1995 *Reflections on Gender and Science* (New Haven, CT: Yale University Press) p 91
[14] Gross P and Levitt N 1994 *Higher Superstition: The Academic Left and Its Quarrels with Science* (Baltimore, MD: Johns Hopkins University Press) ch 3
[15] Sokal A and Bricmont J 1998 *Fashionable Nonsense: Postmodern Intellectuals' Abuse of Science* Intellectual Impostures (London: Profile) ch 10
[16] Sokal A and Bricmont J 1998 *Fashionable Nonsense: Postmodern Intellectuals' Abuse of Science* Intellectual Impostures (London: Profile) ch 6
[17] Sokal A and Bricmont J 1998 *Fashionable Nonsense: Postmodern Intellectuals' Abuse of Science* Intellectual Impostures (London: Profile) ch 8
[18] Gross P and Levitt N 1994 *Higher Superstition: The Academic Left and Its Quarrels with Science* (Baltimore, MD: Johns Hopkins University Press) ch 4
[19] Koertge N (ed) 1998 *A House Built on Sand: Exposing Postmodernist Myths about Science* (Oxford: Oxford University Press) ch 16
[20] Kellermeier J 1995 Mathematics, gender, and culture *Transformations* **6** 35–53
[21] Kellermeier J 1992 Writing word problems that reflect cultural diversity *Transformations* **3** 24–30
[22] Anderson H C 1835–7 *The Emperor's New Clothes. Fairy Tales Told for Children. First Collection* (Copenhagen: Reitzel)
[23] Kuhn T S 1962 *The Structure of Scientific Revolutions* 1st edn (Chicago, IL: University of Chicago Press)
[24] Kastrup H 2007 Ibid
[25] Mallow J 2007 Constructivism in physics education—philosophically problematic, but pedagogically successful *AGORA* **10** 30–2
[26] Hensen *et al* 2015 Loophole-free Bell inequality violation using electron spins separated by 1.3 kilometres *Nature* **526** 682–6
[27] Fensham P, Gunstone R and White R (ed) 1994 *The Content of Science: A Constructivist Approach to Its Teaching and Learning* (London: Falmer)
[28] Østergaard L D, Sillasen M, Hagelskjær J and Bavnhøj H 2010 Inquiry-based science education—does science education in Denmark have use for it? *MONA* **4** 25–43 (in Danish, English translation available by request)
[29] Hake R 1998 Interactive engagement versus traditional methods: a six thousand student survey of mechanics test data for introductory physics courses *Am. J. Phys.* **66** 64–74
[30] Gautreau R and Novemsky L 1997 Concepts first—a small group approach to physics learning *Am. J. Phys.* **65** 418–29
[31] Lorenzo M, Crouch C H and Mazur E 2006 Reducing the gender gap in the physics classroom *Am. J. Phys.* **74** 118–22
[32] McDermott L and Redish E 1999 Resource letter: PER-1. Physics education research *Am. J. Phys.* **67** 755–67 and subsequent work by them and their collaborators
[33] Piaget J 1960 *The Psychology of Intelligence* (London: Routledge)

IOP Publishing

Fear of Physics
And how to help students overcome it
Jeffry V Mallow and Helge Kastrup

Chapter 5

Relations and correlations

In chapter 2 we discussed student attitudes towards science. This included results of questionnaire studies, as well as interviews with students and teachers in various Danish and American institutions. In chapter 3 we discussed students' negative self-statements in the Science Anxiety Clinic. Most of these were attitudes about themselves. Some, however, dealt with attitudes about the nature of science:

- Science is too hard for normal people.
- Only super-brains can understand science.
- Boys are supposed to be good in science.
- Science is not for girls.
- Science is not for me.

Were there more connections? If so, what were they? Did they differ by gender? We examined these through large-scale studies of students taking science classes in the US and Denmark [1]. We did this by combining the attitudes and anxiety questionnaires (chapters 2 and 3 respectively), administering both at the same time. We then looked for correlations between the two. We found a most interesting result. It was not surprising to us, as we were looking for such a connection. But it came out a lot stronger than we had suspected. Science anxiety was correlated with precisely three attitudes that we mentioned in chapter 4 under the rubric of radical constructivism:

- Negativity of science toward the individual.
- Subjective construction of knowledge. ('There are no facts.')
- Inherent bias against women.

These three categories were the same in both cultures. We also saw that the pattern of gender differences in science anxiety has persisted over at least two decades [2, 3]. We found that both science attitudes and science anxiety were correlated with gender. The belief in inherent bias against women was (not surprisingly) significantly stronger among women, and females in both national groups were significantly more science-anxious than males. While correlation is not causation, the perception that science is prejudiced against women may predispose female students toward science anxiety and may motivate them to avoid science[1]. We shall discuss this further in chapter 6. It is the task of educators to change both. If this turns out to be synergistic, so much the better. It makes our job easier.

As we saw in chapter 2, American education students, our future teachers, had some very non-scientific attitudes, and high levels of science anxiety. We wanted to see what the case was for pre-service Danish teachers. We administered our two questionnaires to the group described in chapter 2: student teachers in a Danish teachers' seminary, University College Copenhagen (UCC), training to be lower secondary (middle school or junior high school) teachers in all subjects[2]. We looked for relationships between attitudes and anxiety. We simplified our task by looking for correlations between the three significant attitudes—negativity of science toward the individual, subjective construction of knowledge, and inherent bias against women—and three categories we extracted from the science anxiety questionnaire: calculation (e.g. question 4), experiment (e.g. question 29), and evaluation (e.g. question 42). We found that in the non-science group, females had significantly higher anxiety than males. This was not the case in the sciences, nor in math. However, we found that students, irrespective of gender or field of study, expressed strong negative attitudes towards science. This is a cause for concern, to say the least. In particular, why were the science students negative about science? We can only conjecture that despite their choice of field, they were suspicious of some of the ways that science is practiced and the ways in which some scientific discoveries have been put to use. We can, however, take some encouragement from the fact that they at least will be able to assess scientific claims, and make informed decisions as citizens. This is not the case for the non-science students. They will most likely avoid science and be incapable of assessing the different claims. There is however, a caveat. One can argue that the teacher-training students were those who had chosen a science program, but had not taken the courses yet, so a certain degree of skepticism about their attitudes is to be expected, and of some value. And we don't know how negative they still were, once having finished their studies.

What may we conclude from these UCC studies? First, that science anxiety is present in all groups and, we may be bold enough to say, throughout our countries' populations. Second, that it is gender related for non-science students. Third, that negative attitudes towards science are ubiquitous.

---

[1] However, this is not always the case. We did a small study of Chicago public school science teachers, all female. Some attested to their own science anxiety as students, yet had overcome it.
[2] Note that this is not a precise correlation with the American education students, who were training to be primary school teachers, but the study stands on its own.

A subsequent study of Loyola students showed that science anxiety depended on course of study [4]. Not surprisingly, science students were less anxious than humanities and social science students. Mathematics students fell in between[3]. But in all of the groups, female students' anxieties were significantly higher than those of males. Education and nursing students (primarily female) unfortunately scored very high on the science anxiety scale.

Turning now specifically to physics, connections between attitudes, anxieties, gender, and nationality have to greater or lesser degree been correlated with each other.

A particularly interesting study by Loyola researchers ascertained that physics teaching of general education students lowered their science anxiety [5].

Before closing this chapter, we must emphasize the consequences of the radical constructivist attitudes that we examined in chapter 4. We described social scientists and philosophers who do not see these attitudes as something to overcome, but rather as appropriate descriptions of how things are. The notion that science is in fact inherently biased against women reinforces the reactionary idea that science is inappropriate for women. Two examples of radical gender differentiation are the advice of some Danish school guidance counselors to female students that is their right to refuse to take a course in physics, and the notorious comment (chapter 4) comparing Newton's *Principia* to a rape manual. Similarly, the claim that there are no such things as facts reinforces the reactionary idea of equal validity of all narratives, such as astrology and intelligent design.

There is also an idea prevalent among some post-modern social scientists and philosophers that science can attain some legitimacy if and only if it addresses women's issues in every context. This cannot be universally true. JVM attended a UN-sponsored conference on women, science, and technology [6], in which it was proposed that all science texts needed to show their connection to women's issues. He is co-author of a quantum mechanics text [7]. We cannot see how this can possibly meet that demand.

We believe that science is not inherently cold, or gender biased, or fact-free. We trust that our readers agree.

References

[1] Bryant F, Kastrup H, Udo M, Hislop N, Shefner R and Mallow J 2013 Science anxiety, science attitudes, and constructivism: a binational study *J. Sci. Edu. Technol.* **22** 432–8

[2] Mallow J 1994 Gender-related science anxiety: a first binational study *J. Sci. Edu. Technol.* **3** 227–38

[3] Kastrup H and Mallow J 2016 *Student Attitudes, Student Anxieties and How to Address Them: A Handbook For Science Teachers* (San Rafael, CA/Bristol: Morgan and Claypool/Institute of Physics))

[4] Udo M, Ramsey G P, Reynolds-Alpert S and Mallow J V 2004 Science anxiety and gender in students taking general education science courses *J. Sci. Edu. Technol.* **13** 435–46

---

[3] The most anxious were business students. We have no idea what to make of this.

[5] Udo M, Ramsey G P, Reynolds-Alpert S and Mallow J V 2001 Does physics teaching affect gender-based science anxiety? *J. Sci. Edu. Technol.* **10** 237–47
[6] UNESCO 2010 *Expert Group Meeting: Gender, Science and Technology* (Paris: UN Division for the Advancement of Women and UNESCO) http://www.un.org/womenwatch/daw/egm/gst_2010
[7] Gangopadhyaya A, Mallow J V and Rasinariu C 2017 *Supersymmetric Quantum Mechanics: An Introduction* 2nd edn (Singapore: World Scientific)
[8] Mallow J, Kastrup H, Bryant F, Hislop N, Schefner R and Udo M 2010 Science anxiety, science attitudes, and gender: interviews from a binational study *J. Sci. Edu. Technol.* **19** 356–69

Chapter 6

Gender: not your grandmother's inequity

If you watch the TV series NOVA, you can't help but notice that a large proportion of the top research scientists are women and/or people of color. It is evident that much has changed over the last few decades. But some things have not, or not enough. In this chapter we shall explore both.

6.1 Statistics

In what follows we present data from the US, the UK, and Denmark (DK). We restrict ourselves to university students and their pipeline precursors, secondary school students.

We begin with the pipeline: secondary education. The three educational systems are quite different, but some comparisons are possible.

6.1.1 US

Table 6.1 shows the number of secondary school students taking a physics course for the years 2001, 2009, and 2013.

The number of female physics students continues to grow. Since 2009 females represent about 47% of high-school students taking physics, up from 39% in 1985. Were this the whole story, we would not be very concerned. But 60% of high-school students are girls. Furthermore, as table 6.1 shows, the rate of increase in absolute numbers has stagnated[1].

The sources of this have not been identified, but the radical constructivist paradigm that science is *intrinsically* anti-female (chapter 4) no doubt holds some sway in high school and continues into university. We have seen in our studies described in chapter 5 that these attitudes and the concomitant anxiety are a possible source of the difference. So might be the ongoing societal attitudes towards and by

---

[1] It's possible that this is a one-year glitch. AIP is currently compiling data for subsequent years.

Table 6.1. Number of students taking a high-school course in physics (in thousands) [1].

Year	Male	Female
2001	503	428
2009	716	635
2013	740	636

Table 6.2. Number of students taking A-level physics (in thousands) [2].

Year	Male	Female
2011	20.03	5.17
2016	21.0	5.67

females, independent of radical constructivism. All other things being equal, the percentages[2] might still not be equal. We don't know. But our mission is to make sure that all other things are equal.

6.1.2 UK

The UK secondary school educational system is substantially more complicated than that of the US[3]. The best we do is to obtain A-level numbers, as shown in table 6.2.

Thus only about 20% of students taking A-level physics are girls; this has been essentially flat over the last two decades. The pre-A-level schooling has a strong effect on which girls choose to take physics.

In the report of reference [2] are a series of recommendations to teachers for increasing the number of girls:

1. A senior gender champion is appointed
2. Gender awareness and unconscious bias training is provided for all staff.
3. Sexist language is treated as unacceptable.
4. Use of progression data and formal discussion at the whole-school level.
5. Initiatives are developed that address problems identified in the school data.
6. Subject equity: all subjects are presented to students equally.
7. Careers guidance starts at an early stage.

---

[2] Normalized to the 60% female population in universities.
[3] At least to JVM's American eyes. This was probably a provocation for our having separated from the mother country.

8. Student ownership: students are at the heart of any campaign to tackle gender stereotyping.
   9. Personal, social, health, and economic education includes sessions on equality and diversity.

To quote IOP President Professor Dame Julia Higgins in the foreword to the report, 'It is part of our work to try to understand how boys and girls choose their A-levels, to deconstruct the cultural stereotypes and unconscious bias that discourage girls from taking physics, and to encourage schools to provide girls with the opportunity to study physics at A-level.'

6.2 Biological differences?

In 1980 Camilla Benbow and Julian Stanley claimed, based on statistics that they had gathered, that part of the higher mathematics performance of middle school boys than girls is in part biologically determined [3]. This, as one might imagine, caused a controversy. (Uproar would be more accurate.) Articles arguing for and against the methodology and conclusions abounded. Sheila Tobias[4], who studied the gender gap in math and science at the college level, and whom we shall discuss below and in chapter 11, argued that the Benbow–Stanley sample was skewed [4]. Drawn from an enrichment program run by them for 11-year-old mathematically gifted students, the sample comprised 50% of each. But they did not mention that fewer of the enrollees are girls; others have chosen not to apply because of parental lack of interest and/or because 'the boys are nerdy'.

To be fair, Benbow has never discounted environmental influences on girls' differential performance, although she has maintained the claim that these are still in part biological.

6.3 'It ain't what you don't know that gets you into trouble. It's what you know for sure that just ain't so.' (Mark Twain)

Is science inappropriate for women? Dyed-in-the-wool traditionalists have long answered yes. But as early as the 1980s, radical constructivists have agreed—for different reasons; namely, that science is by its very nature antipathetic to women, and that is why they should and do avoid it. Recall (chapter 4) Sandra Harding's comparing Newton's *Principia* to a rape manual. This became the ruling paradigm.

But do women avoid science? In 1985 chemist and feminist Lilli Hornig demonstrated the opposite. Rather than comparing the large disparity of women versus men in the sciences, she compared the numbers of women in science versus humanities, showing that in fact more women chose the former at all levels [5]. Hornig wrote:

---

[4] Full disclosure: Tobias was a colleague and friend of JVM. She wrote the foreword to the 1986 edition of *Science Anxiety*.

...although it is true that the concentration of women in most science fields is below one-third, compared to about one-half in the humanities, the numbers of women scientists far exceed those of women humanists. Thus, among the total current stock of PhDs in [the US], there are about 63 000 women scientists and about 27 000 women humanists, or a ratio of 2.33. The ratio of new women PhDs in sciences to those in humanities in 1985 stood at 3.44, so that the disparity is growing just as it has among men. The fields regarded as least congenial to women—physical and mathematical sciences—produced over 900 doctorates in 1985, contrasted with about 630 in the so-called traditional fields of English and other modern languages ... more women have been Nobel laureates in the sciences than in either literature or peace endeavors .... When we compare women to men, determining the relative proportions of each sex in various activities, we see great inequalities. When we compare women in one field to those in another, determining how they distribute themselves among the choices open to them, we discover two things: the patterns of choice resemble those of men, and the disadvantages women face are essentially invariant across fields. In short, women face some discrimination in all careers *because they are women*, not because they are unsuited to science or science to them.

Hornig's paper is rarely cited. Despite the data, this paradigm still rules in radical constructivist circles, regrettably trickling down to our students (chapter 5). Our studies (chapter 2) showed that many of them believe in

- Negativity of science toward the individual.
- Subjective construction of knowledge: there are no facts, scientists make things up.
- Inherent bias against women.

Furthermore, these are correlated with science anxiety and its concomitant avoidance (chapters 2 and 5).

6.4 Proportions of females studying science at universities

Hornig is speaking about all the sciences. Although she did note that even in 1985, the physical and mathematical sciences produced over 900 doctorates, contrasted with about 630 in the humanities, the proportion of female physics students is still far below those of chemistry and biology, which have reached approximate equity. Tables 6.3 and 6.4 show the percentages of females taking science courses in the UK and US.

Table 6.3. Female percentages in the US, 2018 [6].

Biology	Chemistry	Physics
60	59	24

Table 6.4. Female percentages in the UK, 2019 [8].

Biology	Chemistry	Physics
63	63	23

While the 24% is way up from about 9% when it was first measured in 1976, it has peaked and levelled in the low 20s since about 2000. In fact, the percentage of women earning bachelor's degrees in physics is declining [7].

While similar to the US physics percentages, this must also be related to the low fraction of female secondary school students taking physics, although it is not obvious to us why this should be so. Only an extended longitudinal study can provide the answers. The results will depend on which and how many of the suggestions listed in section 6.1.2 will be implemented.

The percentages and their levelling off are about the same in Australia, Canada, and Germany [9].

In a series of extended group interviews with upper level undergraduates in Roskilde University's Institute for Mathematics and Physics, more females than males expressed anxiety.

6.5 Seminal research studies

At this point, let us turn to three important research studies that cast light on problems of university science in general and physics in particular [10].

6.5.1 Tobias

The first is the work of Sheila Tobias [11] dating to 1990. We shall discuss this further in chapter 11; here let us remark on her major findings. The book comprises interviews with graduate students with non-science degrees. Each enrolled as an auditor in an undergraduate science course, and reflected on their experiences therein. By and large these were not positive, and in one case a potential science student chose a different path after experiencing a science class. They were put off, inter alia, by the goals of the class ('I want to understand the big questions about physics and the world, but the class is focused on problem-solving.'), the nature of the pedagogy ('The format is entirely lecture.'), and the classroom atmosphere ('There is little interaction between students.'). All of this is still painfully true in many courses despite changes in the intervening decades, such as interactive engagement, group work, and recognition and amelioration of science anxiety.

6.5.2 Seymour and Hewitt

In 1997 Elaine Seymour and Nancy M Hewitt [12] carried out a large-scale study to ascertain why American students transferred from one science major, e.g. physics, to another science major such as chemistry or biology, or left entirely for other fields, such as humanities or the social sciences. In aggregate, 51% of males and 62% of females switched out of a science or engineering major during their university careers. Of these, relatively few switched to another science or mathematics: 48% of males and 59% of females left mathematics, science and engineering altogether. Furthermore, the percentages of males and female students leaving physics were higher than the other sciences. The percentage of females leaving physics was a staggering 72%. Seymour and Hewitt investigated various possibilities, and concluded that those who switched out were as well prepared, as highly motivated, and as good performers as the ones who stayed. In fact, it was the educational experience, the 'culture of science' that was the cause of most of the losses. Switchers of both genders cited poor teaching as one of the top reasons for leaving. This substantiates the earlier work of Tobias [11] regarding putative 'natural' inclinations for science. Furthermore, there were not the gender differences which one might have expected. Women did not leave more often than men because they were attracted to non-science fields. In fact, the difficulty of the science field, pace and workload, low grades, the competitive culture, drove away a substantially greater proportion of males.

A 2019 follow-up study by Seymour and co-workers elaborated on what has and has not changed [13]. Among the still-major culprits is poor teaching, although that has somewhat improved. Part of the improvement was generated by the demonstration that bad teaching led females and students of color leaving in significantly larger numbers that white males. However, much more is to be done, not only in teaching, but in the other areas implicated in the earlier book.

6.5.3 Hake

In 2000 physicist Richard Hake undertook a study of the efficacy of lecturing versus 'interactive engagement' (IE) of students with their teacher and with each other in the classroom [14]. Using the Force Concept Inventory [15] as a pre- and post-test of students' understanding of mechanics, he showed that various forms of IE were superior to straight lecturing. Later he examined the role (if any) of gender on the results [16]. He compared gains ($\langle g \rangle$) in understanding for several universities. While finding different values for the gender component at various of these institutions, he concluded, '...*the $\langle g \rangle$ dependence on the gender "hidden variable" is small relative to the very strong dependence of $\langle g \rangle$ on the degree of interactive engagement (effect size 2.43)...*'. Therefore, in my opinion, '***efforts to move traditional instruction more towards the interactive engagement for ALL students should receive a higher priority than concern for the apparently relatively small gender differences...***' (bold and italics emphasis in original).

6.6 Raising the percentage

What about that still large percentage discrepancy between female and male physics majors? Clearly there is still male bias, but it has been and still is being fought, certainly over the last two decades. Why did the increasing percentage of female physics students grind to a halt? We know from our questionnaires and interviews with

American and Danish physics students that anxiety and the related negative attitudes play a significant role [17]. We do not have data from the UK, but given the similar percentages [18], we would be surprised if it were not a component. That being said, we do not believe that they play as important a role as earlier, and they cannot explain the stagnation of the percentage. We suspect that stereotype threat for females is no longer a major cause, given the numbers of women now studying physics.

One reason for the stalled percentage in the US may be, ironically, the attraction of STEM. Males with very low high-school GPAs in math and science and very low SAT math scores are choosing physics, engineering, and computer science (PECS) just as often as females with much higher math and science achievement [19]. The differences are stark. Male students in the 1st percentile chose PECS at the same rate as females in the 80th percentile. This of course depresses the female percentages. How much of an effect it has is not clear, but the integrated low-and medium-scoring males now choosing PECS constitute a large sample. In addition, while the number of girls taking high-school physics has risen in the US, their percentages decrease in the more advanced courses [20].

Have all of the efforts up to now reached some sort of saturation? Not likely. Not everything has been tried. The American Physical Society's Committee on the Status of Women recommends numerous approaches to increase the number of females (see [7]).

One initiative is the broadening of major options. For example, Loyola offers four majors:

- Biophysics.
- Physics.
- Physics and Computer Science (Physics CS).
- Theoretical Physics and Applied Mathematics (TPAM).

The distribution by gender for the last three years is shown in table 6.5.

Table 6.5. Choices of major by gender, Loyola University.

		F	M
2019	Biophysics	13	13
	Physics	16	36
	Physics CS	0	6
	TPAM	2	11
2020	Biophysics	7	11
	Physics	15	31
	Physics CS	1	8
	TPAM	1	12
2021	Biophysics	13	5
	Physics	15	27
	Physics CS	1	7
	TPAM	3	7

Females are far more likely to choose biophysics over Physics CS and TPAM than are males. But in all cases, they choose pure physics first.

The US average of female physics majors in 2018 was 24% (see table 6.3). The average at Loyola is significantly higher than that. In 2018 it was 36%. We believe an important factor is recruitment of high-school girls. One strategy is to get an invitation to a high school and send a female professor to speak. The concomitant critical mass of female majors leads to minimization of any stereotype threat. Interactions between and among the females likely aids in retention. Loyola's female physics majors met with senior students of both genders during the 2021 fall semester, where they learned about the familial atmosphere of the department and received guidance from them regarding how to succeed in the program. These meetings were facilitated by the female administrative assistant and no faculty members were invited to it.

Finally, group projects (see chapter 10) are required, with careful oversight by the professor of interactions among the group members, especially with regard to gender.

6.7 Where do we go from here?

While this book is designed for the university, we have not hesitated to trace the fear and avoidance of physics to secondary school. But that is not far enough. We need to intervene early, and that means *early*. Chiarelott and Czerniak in the US have shown [21] that as soon as science enters the curriculum, when the students are nine years old, so does gender-based science anxiety—and the topic introduced is typically biology, maybe with some chemistry. This is coupled with teachers' science anxiety, especially with regard to physics, which education school curricula show that they likely have never learned. Our study of attitudes and anxieties of elementary school pre-teachers not only showed a high level of anxiety, but a propensity to believe in pseudoscience [22]. Liaisons between universities and secondary schools are not enough. Somehow, we need to reach out to primary school teachers.

Enough. Let's cut the Gordian knot. That females represent about 47% of US high-school students taking physics is an encouraging statistic. But would anyone accept a number of less than 100% of high-school students taking English? The real fight is to get physics to be required. If some students are anxious, that could be dealt with as we did in the Loyola Science Anxiety Clinic: a physicist and a psychologist working together. Other approaches are possible. But avoidance should not be an option.

References

[1] Number of male and female physics students *AIP* https://www.aip.org/statistics/data-graphics/number-male-and-female-physics-students; Women in physics and astronomy *AIP* https://www.aip.org/statistics/women

[2] IOP 2018 Why not physics? A snapshot of girls' uptake at A-level *Institute of Physics Report* https://www.iop.org/sites/default/files/2018-10/why-not-physics.pdf

[3] Benbow C and Stanley J 1980 Sex differences in mathematical ability: fact or artifact? *Science* **210** 1262–4
[4] Tobias S 1991 Gazette: a newsletter of the Committee on the Status of Women in Physics in the American Physical Society *Am. Phys. Soc.* **11** 6–7
[5] Hornig L 1987 Gender and science *Contributions to the Fourth Gender and Science and Technology Association (GASAT) Conf. (Ann Arbor, MI)*
[6] NCSES 2021 Field of degree: women *NSF* https://ncses.nsf.gov/pubs/nsf21321/report/field-of-degree-women
[7] Women in physics *APS* https://www.aps.org/programs/women/index.cfm
[8] 2019 Why are female students now outnumbering males in A-level science? *STEM Women* https://www.stemwomen.com/blog/2019/08/why-are-female-students-now-outnumbering-males-in-a-level-science
[9] *Women in Physics* International Factsheet https://higherlogicdownload.s3.amazonaws.com/APS/2c0c9f07-6428-4f8e-b9aa-a76098a80cd0/UploadedImages/WiP_InternationalFacts_2020.pdf
[10] Hake R and Mallow J 2008 Gender issues in science/math education (GISME). Available on request
[11] Tobias S 1990 *They're not Dumb, They're Different* (Tucson, AZ: Research Corporation)
[12] Seymour E and Hewitt N 1997 *Talking about Leaving: Why Undergraduates Leave the Sciences* (Boulder, CO: Westview)
[13] Thiry H, Weston T, Harper R, Holland D, Koch A, Drake B, Hunter A and Seymour E 2019 *Talking about Leaving Revisited* (Cham: Springer Nature)
[14] Hake R 1998 Interactive engagement vs traditional methods: a six-thousand student survey of mechanics test data for introductory physics courses *Am. J. Phys.* **66** 64–74
[15] Hestenes D, Wells M and Swackhamer G 1992 Force concept inventory *Phys. Teach.* **30** 141–66
[16] Hake R 2002 Relationship of individual student normalized learning gains in mechanics with gender, high-school physics, and pretest scores on mathematics and spatial visualization *Physics Education Research Conf., Boise, Idaho (August 2002)*
[17] Bryant F, Kastrup H, Udo M, Hislop N, Shefner R and Mallow J 2013 Science anxiety, science attitudes, and constructivism: a binational study *J. Sci. Edu. Technol.* **22** 432–8
[18] Jameson V 2018 Women in physics: why there's a problem and how we can solve it *New Sci.* 7 November https://www.newscientist.com/article/mg24032031-900-women-in-physics-why-theres-a-problem-and-how-we-can-solve-it/#ixzz7TrFAnQzY
[19] Cimpian J, Kim T and Mcdermott Z 2020 Understanding persistent gender gaps in STEM *Science* **376** 1317–9
[20] White S and Langer Tesfaye C 2011 Female students in high school physics *AIP Focus On* https://www.aip.org/sites/default/files/statistics/highschool/hs-studfemale-09.pdf
[21] Chiarelott L and Czerniak C 1985 Science anxiety among elementary school students: an equity issue *J. Educ. Equity Leadership* **5** 291–308
Chiarelott L and Czerniak C 1987 Science anxiety: implications for science curriculum and teaching *Clear. House* **60** 202–5
[22] Kastrup H and Mallow J 2016 *Student Attitudes, Student Anxieties, and How to Address Them: A Handbook for Science Teachers* (London: Institute of Physics)

[23] Mallow J, Kastrup H, Bryant F, Hislop N, Schefner R and Udo M 2010 Science anxiety, science attitudes, and gender: interviews from a binational study *J. Sci. Educ. Technol.* **19** 356–69
[24] Mallow J 1994 Gender-related science anxiety: a first binational study *J. Sci. Educ. Technol.* **3** 227–38
[25] Mallow J 1998 Student attitudes and enrolments in physics, with emphasis on gender, nationality, and science anxiety *Justification and Enrolment Problems in Education Involving Mathematics or Physics* ed J H Jensen, M Niss and T Wedege (Roskilde: Roskilde University Press) pp 237–58
[26] Udo M, Ramsey G, Reynolds-Alpert S and Mallow J 2001 Does physics teaching affect gender-based science anxiety? *J. Sci. Educ. Technol.* **10** 237–47

Chapter 7

Nationality: does it matter?

Is there something special inborn in the Norwegians that makes them better than most other nationalities in cross-country skiing? Is there a special inborn quality in Americans making them better in basketball than most other nations? Not at all. Norway has a tradition and a culture where skiing has been important for centuries. And in most places, good skiing terrain is close. The US has a culture where playing basketball is something done in every school, and many private houses have a basketball net in front of the garage. So, when we talk of national differences, we are not discussing genetic differences between students in Botswana and in Iceland. We are only considering the outcomes of different national traditions, educational systems, and how students are tested. In chapter 3 we compared students in Denmark and in the US. Here we shall start by discussing worldwide comparisons.

7.1 PISA 2015

The acronym PISA means Programme for International Student Assessment. It is a series of investigations carried out in three-year intervals, launched by the Organization for Economic Cooperation and Development (OECD). For each test, groups of 15-year-old adolescents are selected to answer the same questions translated into their mother tongue. The first PISA test was carried out in 2000 with 32 countries. PISA 2022, which was just finished while this was being written, has 85 participating countries. Each one looked at reading, mathematics, and sciences, each time with special emphasis on one of the three. Half of the test time is devoted to the subject in focus and the rest of the time is divided between the two others. For example, in 2000 the focus was on reading, 2003 focused on mathematics, and 2006 on science. We shall consider the 2015 test, the latest focusing on science. Table 7.1 lists the 25 top-rated countries for 2015 for all three subjects and for science. (The US is 31st.)

Table 7.1. Best ratings in math, reading, and science and best ratings in science in PISA 2015. Data adapted from [2].

Average score of PISA 2015 mathematics, science, and reading			PISA 2015 science scores		
1.	Singapore	551.7	1.	Singapore	556
2.	Hong Kong	532.7	2.	Japan	538
3.	Japan	528.7	3.	Estonia	534
4.	Macau	527.3	4.	Taiwan	532
5.	Estonia	524.3	5.	Finland	531
6.	Canada	523.7	6.	Macau	529
7.	Taiwan	523.7	7.	Canada	528
8.	Finland	522.7	8.	Vietnam	525
9.	South Korea	5·19.0	9.	Hong Kong	523
10.	China	5·14.3	10.	China	5·18
11.	Ireland	509.3	11.	South Korea	5·16
12.	Slovenia	509.3	12.	Slovenia	5·13
13.	Germany	508.0	13.	New Zealand	5·13
14.	Netherlands	508.0	14.	Australia	5·10
15.	Switzerland	506.3	15.	Netherlands	509
16.	New Zealand	505.7	16.	Germany	509
17.	Denmark	504.3	17.	United Kingdom	509
18.	Norway	504.3	18.	Switzerland	506
19.	Poland	503.7	19.	Ireland	503
20.	Belgium	502.7	20.	Denmark	502
21.	Australia	502.3	21.	Belgium	502
22.	Vietnam	502.3	22.	Poland	501
23.	United Kingdom	499.7	23.	Portugal	501
24.	Portugal	497.0	24.	Norway	498
25.	France	495.7	25.	United States	496
31.	United States	487.7			

The OECD definition is 'Science literacy is defined as the ability to engage with science-related issues, and with the ideas of science, as a reflective citizen. A scientifically literate person is willing to engage in reasoned discourse about science and technology, which requires the competencies to explain phenomena scientifically, evaluate and design scientific enquiry, and interpret data and evidence scientifically.'

7.1.1 Key features of PISA 2015

Approximately 540 000 students completed the assessment in 2015, representing about 29 million 15-year-olds in the schools of the participating countries and economies.

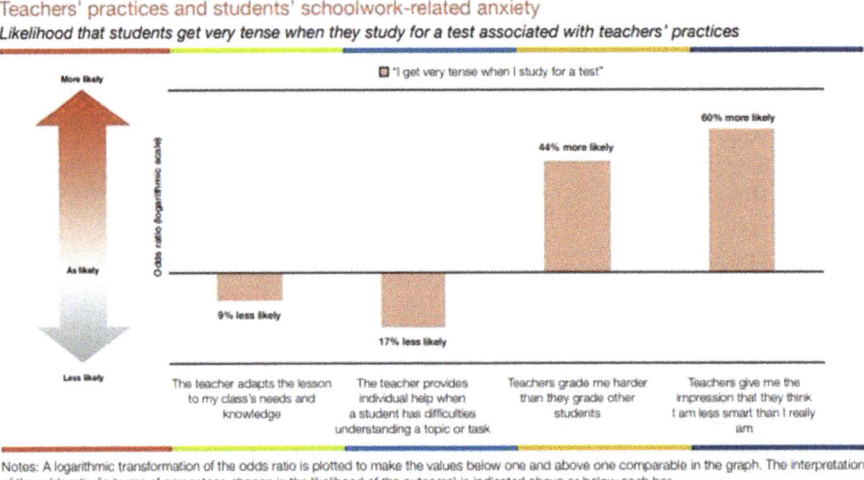

Figure 7.1. Relations between teachers' practices and student anxieties. Reproduced with permission from [3].

Computer-based tests were used, with assessments lasting a total of two hours for each student.

Test items were a mixture of multiple-choice questions and questions requiring students to construct their own responses. The items were organized in groups based on a passage setting out a real-life situation. About 810 min of test items for science, reading, mathematics and collaborative problem solving were covered, with different students taking different combinations of test items.

Students also answered a background questionnaire, which took 35 min to complete. The questionnaire sought information about the students themselves, their homes, and their school and learning experiences. School principals completed a questionnaire that covered the school system and the learning environment. For additional information, some countries/economies decided to distribute a questionnaire to teachers. It was the first time that this optional teacher questionnaire was offered to PISA-participating countries/ economies. Countries could choose two optional questionnaires for students: one asked students about their familiarity with and use of information and communication technologies; and the second sought information about students' education to date, including any interruptions in their schooling, and whether and how they are preparing for a future career. The teachers' pedagogy questionnaires are important, as per the very important graph of the relations between teachers' practices and student anxieties, as in figure 7.1; likewise the statistics on schoolwork related anxiety, figure 7.2.

The students' questionnaires are important because they tell us what attitude and anxiety baggage they may bring to school.

Figure III.4.1 ■ **Prevalence of schoolwork-related anxiety, by gender**
Percentage of students who reported that they "agree" or "strongly agree" with the following statements

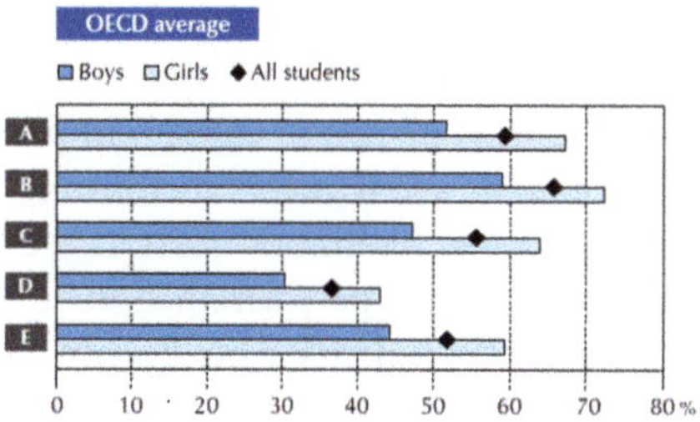

	Index of schoolwork-related anxiety	A	B	C	D	E
Australia	0.2	62	65	68	47	60
Austria	-0.1	64	63	51	19	43
Belgium	-0.2	56	65	42	28	54
Canada	0.2	59	64	64	46	63
Chile	0.1	59	81	56	40	54
Czech Republic	-0.2	55	58	40	32	49
Denmark	0.1	55	65	64	46	54
Estonia	-0.2	51	55	53	28	41
Finland	-0.4	38	44	49	18	37
France	-0.1	62	65	47	29	55
Germany	-0.3	52	53	42	22	35
Greece	-0.1	46	48	59	38	65
Hungary	-0.1	62	66	54	27	46
Iceland	-0.1	48	59	51	37	44
Ireland	0.1	62	69	63	46	55
Israel	-0.3	58	50	44	33	43
Italy	0.5	66	85	70	56	77
Japan	0.3	78	82	62	33	50
Korea	0.1	69	75	55	42	52
Latvia	-0.1	53	68	43	27	47
Luxembourg	-0.2	58	64	48	28	44
Mexico	0.3	72	79	60	50	65
Netherlands	-0.5	34	45	39	14	26
New Zealand	0.3	65	67	72	51	61
Norway	0.1	51	66	61	46	49
Poland	-0.1	62	70	45	26	41
Portugal	0.5	84	88	69	46	65
Slovak Republic	-0.2	61	62	47	29	45
Slovenia	0.1	61	72	62	36	51
Spain	0.4	75	88	67	48	56
Sweden	0.0	56	56	61	41	59
Switzerland	-0.4	48	56	34	21	35
Turkey	0.3	70	74	59	56	69
United Kingdom	0.3	62	67	72	52	55
United States	0.2	63	61	68	43	65

A: I often worry that it will be difficult for me taking a test
B: I worry that I will get poor <grades> at school
C: Even if I am well prepared for a test I feel very anxious
D: I get very tense when I study
E: I get nervous when I don't know how to solve a task at school

OECD average — Boys □ Girls ◆ All students

Figure 7.2. Schoolwork related anxiety. Reproduced with permission from [3].

Seventy-two countries were involved. Singapore was in the top in science, with a mean of 556 while The Dominican Republic was at the bottom, with a mean of 332 this time the mean lying at the 1a level. The difference exceeded two standard deviations.

7.1.2 Gender differences

These were quite interesting:

'In general, boys show greater variation in performance than girls. In all but 18 countries and economies (where the difference is not significant), the variation in science performance (measured by the standard deviation) is larger among boys than among girls.... As a result, on average across OECD countries, the share of top-performing students (those who perform at or above Level 5) is larger among boys than among girls, but so is the share of low-achieving students (those who perform below Level 2 on the science scale). Whereas 8.9% of boys perform at or above Level 5, only 6.5% of girls perform at that level. At the same time, 21.8% of boys do not reach a baseline level of proficiency in science, a slightly larger proportion than that of girls (20.7%)....

'In 33 countries and economies, the share of top performers in science is larger among boys than among girls.... Among the countries where more than 1% of students are top performers in science; e.g., in Austria, Chile, Ireland, Italy, Portugal and Uruguay, around two out of three top-performing students are boys. Finland is the only country in which there are significantly more girls than boys among top performers.

'Boys are over-represented compared to girls among low-achieving students in science in 28 countries/economies, while girls are over-represented in 5 countries/economies .... In the remaining countries/economies, the gender difference in the share of low performers and top performers is not statistically significant.' [2]

7.1.3 Attitudes

The results were generally what one would have guessed:

'In most countries, PISA data show that expectations of future careers in science are positively related to performance in science and, even after accounting for performance, to enjoyment of science activities ... the likelihood that a student expects to pursue a career in science increases as his or her performance in science improves, and this association is positive among both students who do not value science as something particularly interesting and enjoyable (those who are one standard deviation below the OECD average on the index of enjoyment of science) and students who do (those who are one standard deviation above the OECD average on that index). But the association with performance depends on the degree to which students enjoy science.'

Figure 7.1, coupled with a bi-national Danish–American study of correlations between teachers' style (see chapter 2 and 3) and student confidence [1], gives a detailed picture of the correlation of the two.

7.2 The seven levels of proficiency

Figure 7.3 shows the proficiency requirements by which the PISA test is graded. Note that the detailed transformation from Piagetian concrete to formal reasoning model is measured by these levels.

While PISA covers a wide number of countries, we have considered comparisons between Denmark and the US. Our focus was not on performance, but on attitudes and anxiety. These were covered in detail in chapters 2, 3, and 5, and need not be repeated here. In answer to the question of the title of this chapter, the answers were yes and no, depending on which attitudes and anxieties were measured. We repeat here excerpts from these chapters.

Chapter 2. Attitudes

The Danish students described science in the same ways as the American students.

What generalizations can we make from this bi-national study? First, gender plays a major role. In almost every category, the responses of females in both countries differed significantly from those of males. Second, radical constructivist views were articulated by both the Danish and the American students. Third, radical constructivist views about science are linked to science anxiety in both the US and Denmark.

Chapter 3. Anxieties

Here we addressed the responses to the following questions on anxiety:

Level	Lower score limit	Characteristics of tasks
6	708	At Level 6, students can draw on a range of interrelated scientific ideas and concepts from the physical, life and earth and space sciences and use content, procedural and epistemic knowledge in order to offer explanatory hypotheses of novel scientific phenomena, events and processes or to make predictions. In interpreting data and evidence, they are able to discriminate between relevant and irrelevant information and can draw on knowledge external to the normal school curriculum. They can distinguish between arguments that are based on scientific evidence and theory and those based on other considerations. Level 6 students can evaluate competing designs of complex experiments, field studies or simulations and justify their choices.
5	633	At Level 5, students can use abstract scientific ideas or concepts to explain unfamiliar and more complex phenomena, events and processes involving multiple causal links. They are able to apply more sophisticated epistemic knowledge to evaluate alternative experimental designs and justify their choices and use theoretical knowledge to interpret information or make predictions. Level 5 students can evaluate ways of exploring a given question scientifically and identify limitations in interpretations of data sets including sources and the effects of uncertainty in scientific data.
4	559	At Level 4, students can use more complex or more abstract content knowledge, which is either provided or recalled, to construct explanations of more complex or less familiar events and processes. They can conduct experiments involving two or more independent variables in a constrained context. They are able to justify an experimental design, drawing on elements of procedural and epistemic knowledge. Level 4 students can interpret data drawn from a moderately complex data set or less familiar context, draw appropriate conclusions that go beyond the data and provide justifications for their choices.
3	484	At Level 3, students can draw upon moderately complex content knowledge to identify or construct explanations of familiar phenomena. In less familiar or more complex situations, they can construct explanations with relevant cueing or support. They can draw on elements of procedural or epistemic knowledge to carry out a simple experiment in a constrained context. Level 3 students are able to distinguish between scientific and non-scientific issues and identify the evidence supporting a scientific claim.
2	410	At Level 2, students are able to draw on everyday content knowledge and basic procedural knowledge to identify an appropriate scientific explanation, interpret data, and identify the question being addressed in a simple experimental design. They can use basic or everyday scientific knowledge to identify a valid conclusion from a simple data set. Level 2 students demonstrate basic epistemic knowledge by being able to identify questions that can be investigated scientifically.
1a	335	At Level 1a, students are able to use basic or everyday content and procedural knowledge to recognise or identify explanations of simple scientific phenomenon. With support, they can undertake structured scientific enquiries with no more than two variables. They are able to identify simple causal or correlational relationships and interpret graphical and visual data that require a low level of cognitive demand. Level 1a students can select the best scientific explanation for given data in familiar personal, local and global contexts.
1b	261	At Level 1b, students can use basic or everyday scientific knowledge to recognise aspects of familiar or simple phenomenon. They are able to identify simple patterns in data, recognise basic scientific terms and follow explicit instructions to carry out a scientific procedure.

Figure 7.3. The seven levels of proficiency. Reproduced with permission from [2].

What causes of anxiety in science can you identify?
Do you experience anxiety in any subjects or situations?

The simple equation that non-science students would be anxious about science and vice versa did not turn out to be the case. Physics students expressed anxiety about mathematics and chemistry, while the responses of education, humanities, and College Physics students were mixed, with some expressing anxiety about science and some about non-science.

One thing stands out: the influence of teachers. It is they who most affect the anxiety (or lack thereof) of the students.

Chapter 5: Correlations

Science anxiety was correlated with precisely three attitudes of radical constructivism:

1. Negativity of science toward the individual.
2. Subjective construction of knowledge. ('There are no facts.')
3. Inherent bias against women.

These three categories were the same in both cultures. We also saw that the pattern of gender differences in science anxiety has persisted over at least two decades in both cultures. We found that both science attitudes and science anxiety were correlated with gender. The belief in inherent bias against women was (perhaps not surprisingly) significantly stronger among women. Females in both national groups were significantly more science-anxious than males. While correlation is not causation, the perception that science is prejudiced against women may predispose female students toward science anxiety and may motivate them to avoid science.

American education students, our future teachers, had some very non-scientific attitudes (e.g. belief in astrology) and high levels of science anxiety. For the pre-service Danish teachers, we found that in the non-science group, females had significantly higher anxiety than males. This was not the case in the sciences, nor in math. However, we found that students, irrespective of gender or field of study, expressed strong negative attitudes towards science. Why were science students negative about science? We can only conjecture that despite their choice of field, they were suspicious of some of the ways that science is practiced and the ways in which some scientific discoveries have been put to use.

In conclusion, nationality and culture matter, but only in certain areas. And in most cases, they are dwarfed by gender differences.

References

[1] Mallow J V 1995 Students' confidence and teachers' styles: a binational comparison *Am. J. Phys.* **63** 1007–11
[2] OECD 2016 *PISA 2015 Results (Volume I): Excellence and Equity in Education* (Paris: OECD Publishing)
[3] OECD 2017 *PISA 2015 Results (Volume III): Students' Well-Being* (Paris: OECD Publishing)

Chapter 8

Math anxiety: the pipeline choker

8.1 Mathematics: the filter

In the 1970s sociologist Lucy Sells found that of the entering freshmen at the University of California, Berkeley, only 57% of entering males had taken four years of high school mathematics. Only 8% of females had done the same [1].

Sells then looked into the consequences of differing math backgrounds. She found that without four years of high school math, students had effectively barred themselves from the study of the physical sciences, of intermediate statistics (a prerequisite for a career in much of the social science area), and of economics. They had excluded themselves from half of the possible major fields of study at Berkeley, restricting themselves to humanities, fine arts, and the traditionally female 'helping professions', such as guidance counseling, and elementary school teaching. Their options had been foreclosed before they even got to college. Those numbers have not changed over the years as much as one would have hoped. The majority of US English majors, elementary education teachers, and the 'helping professions' such as social workers and nurses are female.

8.1.1 Mathematics and gender

When word about Sells's study began to spread, educators at various schools, especially at women's colleges, realized that there was a serious problem that no one had heretofore considered: math anxiety, and especially its predominance among women.

Educators from the late seventies onward began to question very closely evidence purporting to prove men's natural superiority in math [2]. They were easily able to show that much of the evidence was inconclusive and indeed biased, and that in fact women had been socialized to fear and avoid mathematics[1]. Since this had the effect

---
[1] For more evidence demonstrating that women are not 'naturally inferior' to men in math and science, see chapter 6.

not only of channeling women away from math per se, but also of filtering them completely out of the scientific and technical job markets before they even had a choice in the matter, an immediate remedy was needed. Certainly, long-range education and political action was necessary to change women's self-perceptions and society's view of them as non-mathematical. But in the short run, women already in college needed some assistance to overcome their math anxiety. Thus, math anxiety clinics began to spring up on a number of campuses.

8.1.2 Changes

Slowly but surely, large scale change has occurred. As of 2016, the proportions of US males and females at all levels of high school mathematics were approximately equal. The bad news is that less than half had advanced to pre-calculus [3]. There is a correlation with physics enrollment: 58.5% have taken no physics in high school. (There is also still a gender difference: a somewhat lower completion rate in physics for female graduates compared with that of their male peers: 37% versus 42% [4].) That it is even permissible that so many students are not required to take high school physics is an indictment of the school system. Teachers and administrators apparently have similar attitudes, and likely similar anxieties as the students, but they should not be permitted to devise an anemic curriculum.

8.2 Similarities between math anxiety and science anxiety

It was clear that a poor math background was a barrier to advancement in any technical field, including, of course, the sciences. So there is a connection between doing math and doing science. This may lead us to suspect that math anxiety and science anxiety are closely related to each other. Are they? From our experience with science anxiety, we think the answer is—yes and no. Yes, there are striking similarities between the two anxieties, but no, they are not one and the same.

First let us distinguish between the various sciences in terms of their connection to mathematics. Biology has the least connection (at least biology as taught in the first two years of college). Chemistry and especially physics are more closely related to math, using it daily as a tool for understanding those sciences themselves. Let us say at the outset therefore that people who are anxious about learning biology are not necessarily so because of math anxiety (although we suspect there are biology students who might have become chemistry students or chemistry students who might have become physics students if not for their fear of math). Having said this, let us explore first the similarities between mathematics and the sciences, and then the differences, so that we can distinguish between the anxieties.

The first similarity that strikes a student who attempts a math course or a science course is that both emphasize solving 'word problems': 'If Suzy buys two pounds of Macintosh apples at 79 cents a pound and three pounds of Jonathan apples at 89 cents a pound and makes them into apple sauce, what is the cost per pound of the apple sauce?"[2] These word problems not only occur more and more often in college math courses, but they are the primary test of a student's knowledge. Science word problems are the applications of the principles taught in the science course. Therefore, a student who has become anxious about word problems in high school math courses is likely to carry this anxiety over into both college math and college science. For him or her, math anxiety and science anxiety are indistinguishable.

Science and math skills look very similar when we compare them to the skills necessary for art and literature. Textbooks in math and science are, for example, to be read much more slowly than are humanities texts. Problem solving is peculiar to math and science. Mathematicians and scientists operate under the assumption that nature has an order that it is their task to uncover; humanities scholars, writers, and artists do not make this assumption at all. This difference in approach and in necessary skills leads not only to the myth that there is an 'artistic' and a 'scientific' mind, but also to the myth that those who are talented in math are automatically talented in science, and inversely, those not talented in one are not able to do the other. This myth is as false as the former one. We might use as an example Einstein, who when he first began to conceive of his theory of general relativity, considered to be one of the greatest intellectual achievements in physics, realized he didn't know enough math to work the theory out. He had to go to his friend the mathematician Marcel Grossman and learn the necessary math from him—a task that took several years and that Einstein did not describe as painless. Einstein's physical intuition was superior to his mathematical ability. The two are not one and the same. But many people think they are, especially when compared with the arts and literature. This belief tends to blur the distinction between math and science, and fear of one may quickly produce fear of the other. Both math and science also act as job filters.

The situation in math has its counterpart in the sciences. As we noted earlier, there are about equal numbers of US males and females majoring in biology and chemistry[3], as well as in math, while females constitute less than a quarter of physics majors. This is not to suggest that physicists actively discriminate more than chemists or biologists, but rather that people perceive physics as the hardest, as well (correctly) as the most mathematical of the sciences. Or, some would argue, physics is about inert nature and therefore highly abstract, while chemistry and biology are closer to us as human beings.

Both math and physics anxiety act as gender-selective career filters. A similar selection process operates against disadvantaged minorities. One striking example of this was the case of a black woman in our science anxiety clinic who had been unable to do either math or science since elementary school, when her white teacher had told the class of inner-city youngsters that they could not expect to do well on

---

[2] Let's not even talk about train A leaving San Francisco at 9 AM while train B leaves NY at 11 AM.
[3] This is not however gender equity, since 60% of US college students are female.

standardized tests since they were in competition with suburban whites. Black student proportions selecting to study math and science fall below that of other groups [4]. In a lecture at Loyola, Sylvester J Gates, an African-American physicist who holds the Clark Leadership Chair in Science in the Physics Department at the University of Maryland, pointed out the well-known correlation of mathematical and musical talent, and acerbically asked why this ostensibly fails for Blacks.

8.3 Differences between math anxiety and science anxiety

The first difference is simply a fact that we observed: most students who came to the Loyola Science Anxiety Clinic were usually doing fine in math. They reported little or no anxiety in their math classes, but a great deal of anxiety in studying science. Conversely, there are math-anxious people who can do well in science, although this is more likely in biology than in chemistry or physics. So it is clearly the case that one can be science-anxious without being at all anxious about math. In the clinic we administered both math anxiety and science anxiety questionnaires. There was little correlation between the two. However, those students who did exhibit math as well as science anxiety at the beginning of the clinic were able to reduce both, no doubt because the procedures are similar for the clinics.

One very significant difference is that the theory taught in science lectures derives its validity from results found in the laboratory. Mathematical theories, on the other hand, while true in the same sense that scientific theories are true (testable by independent researchers), are in a sense self-contained: what you see in the math classroom is what you get. What you see in the science classroom is based on something external to theory; namely, the experiment. Thus, an important component of science is absent from the classroom. So in a sense the science classroom suffers from the same drawback as the popular science program on TV: it gives a body of results, while unintentionally concealing the difficult process that went into obtaining them. This is partly rectified by requiring a laboratory course along with the science class—assuming the two are synchronized with each other. This is not to say, of course, that doing creative math is easy, but the way math is done is more closely approximated in the math classroom than the way science is done is approximated in the science classroom.

This difference has a number of consequences. The most obvious is that science anxiety has a component that is absent from math anxiety; namely, laboratory anxiety. Students who may be able to function well in lecture courses may be highly anxious when faced with a morass of equipment and asked to perform an experiment to test the theories that they learned in the class. (See chapter 9 for a more detailed discussion of this.)

Another consequence of the nature of science as contrasted with math is that problem solving in science is viewed by students as somehow different from problem solving in math. Part of this stems from the fact that math is, *grosso modo*, teaching students certain techniques with application, while science is forcing them to apply these techniques by focusing on applications. This is particularly the case for physics.

Let us take the following example. A student is taking a course in calculus and a course in physics. The topic covered in calculus is differentiation. The job of the math teacher is to teach the technique so that the student may apply it not only in pure math, but also in applications, such as in physics. Therefore, once the student has learned, or supposedly learned, how to perform the operation of differentiation, the math test may start by asking two types of questions:

1. Differentiate the function $x = 4at^2$.
2. Find an expression for the velocity of an object whose position varies as the square of the time.

The first of these questions is a straightforward request to the student to demonstrate mastery of the technique. The second question is a word problem involving the student's knowing that velocity is the result of differentiation of position, but it is still an application of the same technique. The student who is not math-anxious will recognize this by the context and will probably be able to work the second as well as the first problem.

Now consider the same student in a physics course. Here, the topic is kinematics. This student will not be given question 1. When given a problem similar or identical to question 2, but in the context of a physics exam, he or she may be too anxious to do it. How is that possible? The physics course is drawing not on a limited area of abstract knowledge, such as differentiation, but is instead using differentiation to explain the results of physical experience. So the student first of all does not get as good a clue from context about what is being asked for, since skills other than differentiation are needed to describe motion.

Second, and very important, the student perceives the math course as covering separate, well-compartmentalized topics, while he or she perceives the physics course as describing the physical Universe using mathematical models. The science-anxious physics student assumes that there is something about nature that others have grasped but that he or she has not. The student goes looking for the forest while ignoring the trees, and consequently fails to solve the problem. This familiar feeling that 'something is missing' characterizes the science-anxious student who may not be math-anxious. One of the leaders in math anxiety reduction, Sheila Tobias, described a related phenomenon in math anxiety; namely, the focusing on irrelevant details as a way to avoid doing the math [5]. To the purely science-anxious student, math usually seems free of such irrelevant detail, but he or she is always suspicious that science contains important details that he or she missed while trying to understand the physical model of nature that has been presented. It is essential to realize that this perception about science is false. The science-anxious student tends to read more into the content of science than there may be: if a question on an exam looks easy, then the student assumes that he or she has missed the point. This is not to suggest that science is easy or one-dimensional. Certainly, as one progresses through the study of science, one sees more and more facets of nature, and indeed one begins to see the forest. However, this is not something that is immediately evident. It is a process requiring practice. But the science-anxious student seems to assume that everyone else is immediately grasping the 'big picture'. This erroneous belief lies at the heart of much of the fear of science.

Three further differences between math and science, each of which can be a source of anxiety, are:

- *Orders of magnitude.* Both in math and in physics contexts, this causes anxiety if the student cannot tell if the answer is within reason or not. For example, in math—say, making a polynomial division, and in the sciences—say, computing the impedance of an electrical circuit, can produce anxiety. But the sciences have an extra difficulty: orders of magnitude. For example, what is the mass of a globular cluster? Is it 10^{30} kg or 10^{36} kg? What is the size of a uranium nucleus? Is it 12×10^{-9} m or is it 12×10^{-15} m? Students are not professional physicists. They can have no sense of which of these makes sense.
- *Units.* Many scientific units are frighteningly unfamiliar and strange. What are the units of the gravitational constant, of Coulomb's constant, or Planck's constant? How do they change from one unit system to another? And even in a relatively simple unit like watts = kg m^2 s^{-3}, why seconds to the minus third power? Math does not normally have such problems.
- *Dangers.* Laboratory security is important. There are good reasons to worry about poisonous chemicals and biological materials, about electrical currents and ionizing radiation. These scary dangers, or the fear of them even when the laboratory is safe, are nonexistent in the math class.

We can complete our consideration of the differences between math and science anxieties by asking, 'How do people view the relevance of math and science in their lives?' It is clear that they view math as more immediate and relevant. Math anxiety includes such issues as inability to balance a checkbook, or add up the cost of groceries, or figure out your income tax. Anyone who has gone to school has had contact with math. It is familiar, whether they like it or not. Not so for science. It is possible, thanks to past generations of science-anxious educators, for students to obtain high school diplomas despite minimal contact with it. Science is omnipresent, and in a more threatening way than math is: an unbalanced checkbook is not as dangerous as an unbalanced nuclear reactor. Fear of the applications of science seems to produce a fear of learning science, which places people even more at the mercy of science and its technological offshoots. Students in both high school and college shy away from science, especially physics, at least as much and probably more than away from math. Despite the large advance from 8% in the 1970s to a much closer balance today, avoidance of high school math and its related anxiety choke the pipeline before students, both male and female, even get to college. The blame lies largely with the national standards for secondary schools. Higher level math courses should not be optional. To this we might add that the common situation in the US, where the only science requirement is biology, also needs to change.

References

[1] Sells L 1978 Mathematics as a critical filter *Sci. Teach.* **45** 28–9
[2] Benbow C P and Stanley J 1980 Sex differences in mathematical ability: fact or artifact? *Science* **210** 1262–4
See for example; Benbow C P and Stanley J 1983 More facts *Science* **222** 1029–31
Benbow C P and Lubinsk D 1993 Psychological profiles of the mathematically talented: some sex differences and evidence supporting their biological basis *The Origins and Development of High Ability. Ciba Foundation Symp.* vol 178 (Oxford: Wiley) pp 44–66
[3] National Science Board 2018 Elementary and secondary mathematics and science education *Science and Engineering Indicators 2018* https://nsf.gov/statistics/2018/nsb20181/report/sections/elementary-and-secondary-mathematics-and-science-education/high-school-coursetaking-in-mathematics-and-science
[4] National Center for Education Statistics 2022 High school mathematics and science course completion *Condition of Education* https://nces.ed.gov/programs/coe/indicator/sod/high-school-courses
[5] Tobias S 1978 *Overcoming Math Anxiety* (New York: Norton)

Chapter 9

Laboratory pedagogy: you can't get it from a cookbook

We know from the our research that a key aspect of science anxiety is fear of laboratory experiments. And we know that many students getting anxious in front of, for example, a gas burner or a microscope, do not manifest equivalent anxiety in front of a camping gas stove or a complicated camera.

On the other hand, experimental experience is a key to learning science, for several reasons:

- Experimental work can be fun.
- Science is about building models of the world, such as *inter alia* relativity, quantum mechanics, plate tectonics, evolution of species, the periodic table of elements. All of these grew out of a never-ending series of experiments and observations. It is important for students to understand this process. One way of learning that is by doing more and more complicated experimental work themselves.
- The fascination of science is fostered through Aha!-experiences: seeing it for yourself is much stronger than being told. Looking through a simple pair of binoculars at the rings of Saturn is stronger than viewing a good observatory photo. (And it makes it obvious why Galilei with his primitive telescope called it 'the planet with ears').
- Experimental work is a good teacher of the limitations of what we can measure. (Why can't a light microscope see objects smaller than 500 nm?)
- Experimental work gives a good working knowledge of units, orders of magnitude, and factors of ten.
- Experimental work gives knowledge of security, such as chemical, electrical, nuclear precaution measures.
- Many concepts introduced in theoretical lectures are very abstract: think of for instance buoyance, conductivity, refraction and reflection, chemical

valence, half-life, standing waves etc. But watching phenomena often helps. For instance, seeing a vibrating guitar string with a stroboscopic light source immediately helps one understand the concept of a standing wave.
- For those students wishing to work as scientists, laboratory technicians, nurses, doctors, and many other vocations, experimental skills are necessary. The same goes for normal family-life, household-members, and citizens.
- Experimental work is gender-equalizing if carried out with care. As we noted earlier, traditionally mixed-gender groups had the male participants touching and using the equipment, and females taking notes, i.e. acting as secretaries. Fortunately, these roles are slowly changing, but we are still far from real equality [1], [2]. One—perhaps some would think drastic—means is to form the working groups according to gender[1]. Whether one uses single-gender groups or not, is it important to carefully specify individual tasks for group members, so that everyone gets a chance to handle equipment, draw graphs, propound hypotheses, and formulate interim and final conclusions (HK has used Kagan's concept of cooperative learning [3] in forming physics working groups.)
- Experimental work is a good alternative or supplement to classroom lecturing. But experimental demonstrations in class are also important, as are gedanken experiments.

Here is a parody of a laboratory exercise, shown in figure 9.1.

9.1 The Archimedean law of buoyancy, version 1

Figure 9.1 shows an experimental set-up for the 'cookbook' version of measurement of the Archimedean buoyancy law.

1. Take one of the AB-boxes from cupboard 12-4.
2. Check that it contains three metal cylinders made of aluminum, brass, and lead respectively.
3. Weigh each of them.
4. With all three cylinders:
    a. Hang the cylinder on a piece of string.
    b. Place a measuring glass with water directly under the cylinder.
    c. Measure the mass of glass plus water and read the volume of water.
    d. Lower the cylinder so that it is covered with water.
    e. Measure the volume of water with the cylinder.
    f. Make a second reading of the weight.

---

[1] The word *gender* is a word to be used with care. How many genders are there: are non-binary persons, cis-genders, trans persons, eunuchs... separate genders. In this book we speak traditionally of female and male genders, for reasons of brevity, hoping not to hurt the feelings of readers regarding themselves as something else.

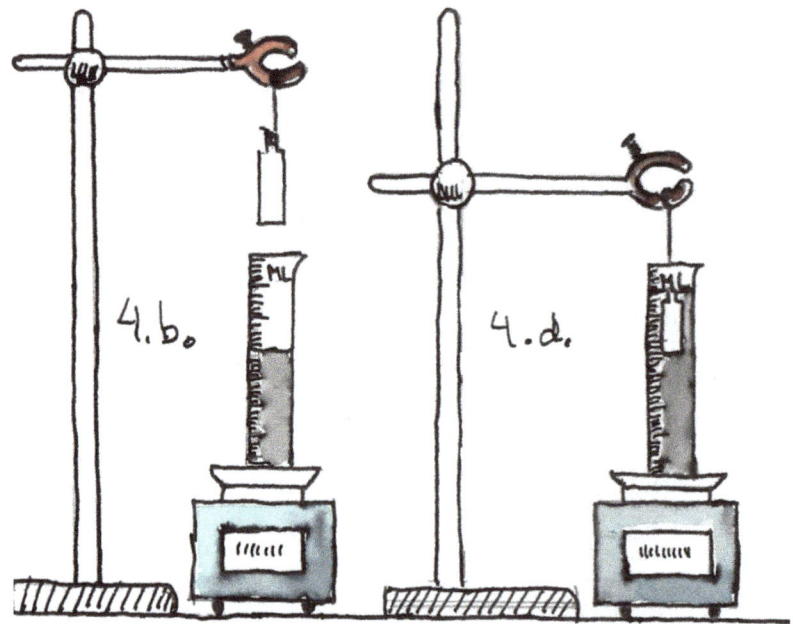

Figure 9.1. Experimental set-up for the Archimedean law of buoyancy, version 1.

Table 9.1. A 'cookbook' experiment.

M_{Al}	$M_{glass + water}$	V_{water}	$V_{cylinder + water}$	$V_{cylinder}$	$\rho_{Al} = M_{Al}/V_{cylinder}$	B
M_{Brass}	$M_{glass + water}$	V_{water}	$V_{cylinder + water}$	$V_{cylinder}$	$\rho_{Brass} = M_{Brass}/V_{cylinder}$	B
M_{Pb}	$M_{glass + water}$	V_{water}	$V_{cylinder + water}$	$V_{cylinder}$	$\rho_{Pb} = M_{Pb}/V_{cylinder}$	B
M_{Al}	$M_{glass + alcohol}$	$V_{alcohol}$	$V_{cylinder + alcohol}$	$V_{cylinder}$	$\rho_{Al} = M_{Al}/V_{cylinder}$	B
M_{Brass}	$M_{glass + alcohol}$	$V_{alcohol}$	$V_{cylinder + alcohol}$	$V_{cylinder}$	$\rho_{Brass} = M_{Brass}/V_{cylinder}$	B
M_{Pb}	$M_{glass + alcohol}$	$V_{alcohol}$	$V_{cylinder + alcohol}$	$V_{cylinder}$	$\rho_{Pb} = M_{Pb}/V_{cylinder}$	B

      g. Compute the density of the metal of the cylinder, ρ.
      h. What is the buoyancy measured? Compare with the volume of the cylinder.
  5. Repeat sequence 4 with denatured alcohol.
  6. Fill in your measurements in the form below.
  7. Does your lab work confirm the Archimedean law?

Details of the 'cookbook' experiment are shown in table 9.1.

Why is this a parody?[2] We consider 'number of freedoms', a pedagogic analogue to mathematical degrees of freedom. This concept was introduced by

---

[2] When HK started teaching physics and chemistry back in the 1960s, his students carried out many labs like the one shown.

Björn Andersson [4]. The cookbook version of the exercise has zero. As the following chart shows, the number increases based on how much freedom of choice is given in the parameters shown. The nature of the freedom can vary.

Number of freedoms	Problem	Procedure	Answer
0 (cookbook)	Given	Given	Given
1	Given	Given	Open
2	Given	Open	Open
3	Open	Open	Open

Today it is generally accepted that more than the three mentioned numbers of freedoms exist. One is the choice of formation of the working group. Who? How many? Single or mixed genders? What are the roles of the individual members? Another is how the results are to be presented: A written report? A PowerPoint presentation? A YouTube video? A song? ...

Labs with no number of freedoms, such as Archimedes version 1 above, can easily be carried out without any learning taking place, and without raising interest or even enthusiasm for the subject matter. They won't urge the student to think out of the box.

There is nothing wrong with the Archimedean experiment. It will work fine as a teacher demonstration, especially if you can somehow project the reading from the digital weight on the screen, so everybody can see. And the somewhat surprising result that the buoyancy force is directly registered on the digital output will demonstrate its full effect.

9.2 The Archimedean law of buoyancy, version 2

1. Discuss the law of Archimedes.
2. Could you use it to measure the volume of a given cylinder?
3. Try to design an experiment to measure the density of a metal cylinder. (You can find such cylinders in cupboard 12-4.)
4. How will you describe/record your experiment for the rest of the class? Do so.
5. Consider how the Archimedean law works with different fluids.
6. Design an experiment showing your thoughts. Do so.

In this version, the number of freedoms is at least two: the freedom of procedure and the freedom of presentation. One might argue that there is also an element of freedom in the answer found.

One way of introducing numbers of freedom is through STEM[3] (Science, Technology, Engineering, Mathematics). Christine M Cunningham [5] gives the following arguments for STEM education as early as in primary school:

---
[3] STEM was earlier often called SMET.

Figure 9.2. The Archimedean law of buoyancy version 3.

- Engineering helps children understand and improve their world.
- Engineering fosters problem-solving skills and dispositions.
- Engineering can increase motivation, engagement, responsibility, and agency for learning.
- Engineering can improve math and science achievement.
- Engineering can improve access to stem careers.
- Engineering improves educational equity.
- Engineering has the potential to transform instruction.
- Engineering is included in state and national standards.

When Christine M Cunningham speaks of STEM education, she refers to all of the natural sciences—biology, chemistry, physics, geography, etc.

The Archimedean law of buoyancy 2 above is a simple example of asking the students to design an experiment and perform it. One might even argue that it is merely a Pinocchian way of making students 'guess what the teacher thinks'. That is reinventing the Archimedean law of buoyancy 1. So let us instead look at figure 9.2.

9.3 The Archimedean law of buoyancy, version 3

1. Build a simple boat (toy-size). You can use a 3D program such as SketchUp for modeling and computations.
2. Compute how deep it will sink in water.
3. Try to compute how deep it will sink loaded with different masses.
4. Mark the different result on the outside of the boat.
5. Put the ship in water and vary your theoretical waterlines.

6. Try the same exercise with salt water[4].
7. Look up the words 'Plimsoll line' on the Internet. (You could start with https://en.wikipedia.org/wiki/Waterline.)

Version 3 is a simple STEM version. It could easily be extended to encompass more physics.

Conclusion: The more freedom you give your students in their labs—the more numbers of freedom—the better. Remember you are there to help them. Don't try to make them play attached to your strings. Encourage them to misinterpret your rules when solving the assignment by lateral thinking.

9.4 Labs and IT

IT and science teaching will be the subject of chapter 13 in this book. But here are a few key points:

- With easily available data logger equipment, students can do what is done in scientific laboratories. Some of these are available in tablet/mobile phone versions. Advantages are that many consecutive measurements, measurements done on short time scales and distances, or measurements done regularly over a longer time scale such as 24 h in a week are possible.
- Many scientific models can be illustrated with simulation apps normally called applets. Some of these are available in tablet/mobile phone versions.
- The data loggers can register measurements directly, in spreadsheets, mathematical programs, graphic programs, and others.
- Students can make mathematical calculations normally too difficult or too time-consuming to solve by pencil and paper, with computer programs such as GeoGebra, Mathematica, Maple, and F-pro.
- The Internet is a rich source of information. But as it is generally not peer-reviewed, it must be used with great caution. Teachers should show examples of pitfalls, such as misinformation.

References

[1] Beyer K 1991 Gender, science anxiety and learning style *Contributions to the 6th GASAT Conf. (Melbourne, Australia)*
Beyer K, Blegaa S and Vedelsby M 1985 Sex-roles and physics education *Contributions to the 3rd GASAT Conf. (London, UK)*
Beyer K and Reich J 1987 Why are many girls inhibited from learning scientific concepts in physics? *Contributions to the 4th GASAT Conf. (Ann Arbor, MI)*
Beyer K and Vedelsby M 1983 Girls and physics—a Danish project *Contributions to the 2nd GASAT Conf. (Oslo, Norway)*
[2] Mallow J and Kastrup H 2016 *Student Attitudes, Student Anxieties, and How to Address Them: A Handbook for Science Teachers* (London: Institute of Physics)

---

[4] If you are close to the sea use seawater, otherwise make your own brine.

[3] Kagan S and Kagan M 2009 *Cooperative Learning* (San Clemente, CA: Kagan Cooperative Learning)
[4] Andersson B 1989 *Grundskolans Naturvetenskab* (Stockholm: Educational Publishers)
[5] Cunningham C 2018 *Engineering in Elementary STEM Education Curriculum Design, Instruction, Learning, and Assessment* (New York: Museum of Science)

IOP Publishing

Fear of Physics
And how to help students overcome it
Jeffry V Mallow and Helge Kastrup

Chapter 10

Group project pedagogy: more heads than one

In earlier chapters we have mentioned group projects both as a part of courses or on its own. This is one of the innovative approaches to science education, a model of interactive engagement [1, 2]. Group work has been shown to enhance student performance, as well as to improve retention in the major and at the university [3–8]. There is also an important and increasingly acknowledged gender component in project work. Females report that they prefer group projects to traditional lectures, because of the interactive, cooperative components and the control of individual competition [9–11]. In fact, projects can and should be designed to build student confidence in both females and males.

10.1 Some characteristics of group project work

We define a group project as an assignment in which a group of students is asked to solve or analyze a given question that is carefully planned to achieve a particular aim. The planning is as important as the aim. Key words are planning, sharing, listening, empathy, and respect.

Working in groups is a preparation for how scientists work much of the time. Today, research is regularly performed by scientists on different continents communicating electronically. Many papers are co-authored by researchers without their meeting physically[1]. Nevertheless, it is group work. When you ask the participants, they are often not able tell who wrote what[2].

Often the expected outcome of a group project is not given beforehand. There is no clearly defined border between a group project and research. But the

---

[1] This book has been written by two close friends who have formerly met many times but now communicate electronically on a daily basis, as one lives in Illinois and the other in Denmark. Because of COVID-19 we have regrettably not been sitting at the same table for several years.
[2] This is also the case with us. We can tell who wrote the first draft of a chapter, but thereafter it passes over the Atlantic Ocean many times, changing each time. This is part of the fun!

purpose of working in groups is not only to teach students how to do research. It is valuable training for becoming an educated citizen. For example, many tasks are solved in groups in the workplace, in setting up a demonstration or preparing a conference.

A project has to have a number of degrees of freedom (see chapter 9). If you prescribe in too much detail how the students are going to work in a group, you remove a degree of freedom.

Working in a group can help an anxious student relax. Tasks are shared with peers. Group members can test ideas before they are written down. Responsibility for using complicated equipment is easier when shared.

But group members may be prone to allowing the dominant member to make the decisions, use the best equipment, and get most of the credit. As a consequence, the least dominant group members will quickly lose interest and will do, say, and learn only little. This has been mostly characteristic of male groups. Often female groups are better at sharing decisions, planning, and responsibilities [9].

As we have noted several times, working in mixed gender groups has traditionally allowed the males to make the decisions and perform the experimental part while the females act as secretaries. We see this gender difference diminishing in the Western world, but teachers should be aware of the risk. An opposing risk is to encourage single-gender groups, which might allow females to remain in their comfort zone. But in the long run, this is not a good idea. For one thing, it is not the real world; for another, it allows all-male groups to do the same, a very bad idea. The analogy would be racial segregation. It can also, for women and some minorities, provoke stereotype threat: a risk of confirming as self-characteristic, a negative stereotype about one's group [12]. It correlates with underperformance on tests. It also strongly correlates with anxiety [13].

The pedagogical basis for project work and numerous other educational practices owes much of its theoretical basis to the work of Lev Vygotsky, Jean Piaget, and Jerome Brunner. Section 10.4 describes their contributions.

10.2 Examples of projects: Denmark upper secondary schools

Here we describe projects for typically three to five students, for a duration of one to two lessons/hours. The first two can be done without access to a lab.

10.2.1 Small group projects

1. The group is given this piece of information: 'If you stayed in the same place for a year, looking in one direction you would observe the Sun revolve around you 365.24 times, but you would observe the stars revolve around you 366.24 times.' They are asked to explain why there is exactly the difference of 1.00 between the two numbers and to create a model/video/demonstration/PowerPoint slide show explaining the fact to the rest of the class.

Figure 10.1. Light from an incandescent lamp seen through a CD without metal.

2. The group is given this piece of information: 'The orbit of the Moon around the Earth and the orbit of the Earth around the Sun are not exactly in the same plane. Why do we not have one solar eclipse and one lunar eclipse every 28 days?' They are asked to explain this and to create a model/video/demonstration/PowerPoint slide show explaining the fact to the rest of the class.
3. The students are given two meters of copper wire and are told to measure the resistance of the wire and copper's resistivity and conductance[3] in at least two different ways. If the concepts were not already discussed in class, they must find the definitions for themselves.
4. The students are given a CD without the metal covering[4] and are asked to look through it at different light sources such as fluorescent lamps, LED lamps, and incandescent lamps, and to observe the different spectra, (perhaps with the help of the Internet) to explain the differences. See figures 10.1–10.3. Students are asked to create a model/video/demonstration/PowerPoint slide show to explain what they have found to the rest of the class.

---

[3] HK used to give his students lacquered copper wire without telling them. They often came back and claimed that the wire had no resistance (really meaning that it had infinite resistance) till they were told to use a knife to scrape off the varnish.

[4] Normally found in packs of CDs to burn. (You can see the light mirrored in an already-burned CD.)

Figure 10.2. Light from a fluorescent lamp seen through a CD without metal.

Figure 10.3. Light from a fluorescent lamp mirrored in a normal CD.

10.2.2 Large group projects

We here give examples of projects for typically two to five students for a duration of several days. On the mathematics–physics line at Kildegaard Gymnasium we had a theme week divided equally between physics and mathematics. Each had five half-days working with a theme chosen by them. In physics these themes were always experimental, and the group rapport was part of curriculum for the oral examinations at the end of term. The students worked independently in groups, but the teacher was present most of the time.

1. Build a Savonius rotor out of an oil barrel cut in two. Try to measure how much power it can deliver for different wind speeds.
2. Build a solar collector out of a radiator painted black. Try to measure how much power it can deliver for sunlight hitting it at different angles.
3. Use a food calorimeter to measure the energy content in different food groups. Discuss what that means for one's daily intake.

10.3 University group projects

Of the various university group project models, we shall focus on two in which JVM has participated. The first was at Loyola University Chicago (LUC). The second was at Roskilde University Center (RUC) in Denmark [14].

10.3.1 LUC projects

Second semester physics majors at Loyola University enrol in a 'Freshman Group Project' course. Begun several decades ago, this is a required one-credit course. Instructors are generally present for each group meeting, approximately once a week.

10.3.1.1 Sample list of freshman projects
1. The inverse Feynman sprinkler.
2. High friction surfaces.
3. Study of a rolling sphere on a curved track.
4. Dynamics of motion in vertical circles.
5. Seismological models in the laboratory.
6. Parameters controlling projectile trajectories.
7. The effect of drag force upon design and propulsion.
8. Friction on ice.
9. Anharmonic oscillators.
10. Rolling oscillators.

10.3.1.2 Sample projects: student descriptions
1. *Analysis of pianos.* The purpose of our project is to understand how string vibrational and acoustical properties differ in grand and upright pianos. To study string vibrations, we built a piano string holder and hammer mechanism to compare free vibrations to those in a grand piano. We used high-speed digital photography to analyze the string vibrations. We observed multiple oscillation modes in the video data and were able to reproduce their properties in Mathematica. To analyze the acoustical properties of pianos, we sampled multiple octaves of notes, providing a wide range of data for comparison. The notes were played with different forces on the key to

measure the effect of impulse on string vibration. The sound data allowed us to compare the acoustical properties of the grand and upright pianos.
2. *Optical response to changing water droplet shape.* Use of water droplets to focus light into solar cells requires a consistent focal length. Droplets of varying sizes were studied and the size, height, and focal length were recorded. By modeling the droplets, we found that small droplets were basically spherical in shape while large droplets were subject to a significant deformation. By modeling the deformed droplets as having a dome shape, we found a consistent height of the spherical portion of the droplet independent of the droplet size. This results in large variations in focal length with droplet size and negates the use of water droplets as lenses for solar cells unless the size of the droplet can be tightly controlled.
3. *Behavior of a loaded string.* This experiment studies the behavior of standing waves on a 'loaded' string with one or more masses attached. We wanted to see the difference between these waves and the waves of an unloaded string. We expected waves with nodes occurring where the masses were placed to be the same, while other waves would have different shapes and frequencies. The experiment is of interest because a string with periodically placed masses serves as a good mechanical analog for the movement of electrons through a metal.
4. *Preliminary investigation of wind turbines and solar cells.* We were interested in sustainable energy generation. We investigated the power output and efficiency in simple drag-type wind turbines for turbines with constant total area, constant blade number, and for various angles of blade orientation for each situation. For solar energy, we used the 'variable load method' to find the voltages and currents at various loads, which were used to find the power output and various parameters.
5. *Study of large amplitude oscillations.* Simple harmonic motion is of fundamental importance in physics. If the angular amplitude of oscillation is small enough, less than 5°, the motion of a simple or physical pendulum can be safely assumed to be simple harmonic, and the time period of oscillation ought to be independent of the amplitude. We studied the motion of a simple pendulum for large amplitudes. We also designed and studied a physical pendulum using a bicycle wheel with a mass attached to the inner rim of the wheel. We compared our observations with numerically obtained results.

Final presentations are twelve minutes long, followed by 3 min of questions posed by both faculty and fellow students. Grading was individual.

This activity has become a valued part of our physics majors' program, and is enthusiastically supported by both students and faculty. Since faculty are involved as project advisors, students come to know all of them, whether or not they have been or will be their classroom instructors. They also acquire early experience with research design and activity, as well as with written and oral presentations. Exit interviews with graduating seniors have confirmed the unanimous student enthusiasm for group project work.

In some cases, students continued and expanded their projects into subsequent years. A number of them resulted in publications in, *inter alia*, *The Physics Teacher*, the *American Journal of Physics*, and the *European Journal of Physics*.

10.3.2 RUC projects

From its inception in 1972, Roskilde University, in suburban Copenhagen, embarked on a radical experiment in pedagogy. Its founders, themselves products of the student revolutions of the late sixties, decided on a revolutionary curriculum: 50% course work and 50% group research projects. This model, with some modifications, has been in existence for almost fifty years, and has proved its value both in external measures of student learning, and in preparing students for careers. External evaluators from other universities consistently rate RUC students high in knowledge of both content and methodology of their chosen academic fields. Corporations and employers seek out RUC graduates for their group interaction skills and their independent problem-solving abilities[5]. RUC's model of group research has been copied by other Danish and European universities, although not necessarily at the scale of 50% of the academic program.

Project work is based on the principle of exemplarity: the idea that one can learn a subject by a detailed study of one aspect of it. One justification for the focus on exemplarity has been political: a holistic society requires a pedagogy that mirrors 'real life'; in particular, that dispenses with a (presumably) artificial disciplinarity, substituting instead the interaction of various disciplines. A second justification for exemplarity is that the sheer volume of knowledge in any particular discipline is so massive that only the selection of a part for detailed study is capable of providing students with mastery.

The RUC curriculum is divided into two parts. Basic Studies comprises the first two years. Four projects are required, one project per semester. Advanced Studies comprises subsequent years[6]. Some of the final project reports are shorter than 30 pages; the average is 60–100 pages.

10.3.2.1 Group project work: Basic Studies
Here is a list of recent physics- and mathematics-based Basic Studies projects. Note the disciplinary breadth of these projects, which comprise hands-on experiments, theoretical studies, philosophy of science, science education, and science and society. At least one of the four must be a hands-on experiment. (By contrast, almost all of the LUC freshman projects have an experimental component.)

1. Application of science in technology and society.
2. Biomass versus greenhouse gases.
3. Death rays? (electromagnetic waves and cancer)

---

[5] While this last one may not be the precise outcome hoped for by the founders of RUC, many of them Marxists, it is nevertheless touted by RUC officials and faculty when helping their students find employment in the public and private sectors.
[6] In some cases into Master's degree programs.

4. Models, theories, experiments.
5. Near-infrared spectroscopy.
6. Theories of climate change.
7. Simulation of crater formation.
8. Reflection upon science and explanation/communication of science.
9. Experts: a study of scientific dishonesty.
10. Time: a physical–philosophical report of time's status.

10.3.2.2 Group project work: Advanced Studies
Upon completion of their two years of Basic Studies, students select their major(s) and then join one of the ten institutes in which their chosen field or fields reside. Physics and mathematics students affiliate with the Institute for Mathematics and Physics Education, Research, and Applications (IMFUFA). Project topic areas are divided into three categories: Metaprojects—philosophy, history, and sociology of science, theories of knowledge, science pedagogy. Toning—applications of science or communication/dissemination of science. Internal—pure science, interaction of theory and experiment. Both Metaprojects and Tonings are reconfigurations and outgrowths of Basic Studies semesters 1 and 3: application, reflection, and communication. Internal projects are analogous to Basic Studies semester 2: models, theories, and experiments. Projects are also subdivided into two time/credit groups: Small Projects—one semester; and Large Projects—two semesters. Students desiring to complete the Master's Degree must do a project in each of the three categories; a Large Project will become their master's thesis. (Bachelor's Degree requirements are somewhat lower.) Below are listed recent physics projects completed by students in the IMFUFA Institute. Note the flexibility within a single institute and between institutes. 'Modeling of a windmill wing' falls under both 'Toning: applications of physics' and 'Internal (pure) physics'. 'The power of illustration' is an interdisciplinary Master's thesis in Physics and Communications. Here is a list of recent Large Projects of the IMFUFA Institute.

1. Knowledge and quantum mechanics (Metaproject: 1/2 year).
2. The crucial experiment (Metaproject: 1/2 year).
3. The hydraulic spring (Internal: 1 year).
4. Expansion of the Universe: history and theory of science (Metaproject: 1/2 year).
5. Physics in secondary school: a schematic overview of students' and teachers' opinions on experiments in physics and chemistry (Metaproject: 1/2 year).
6. Viscoelasticity: determination of a material's viscoelastic characteristics from the material's reflection of sound waves (Toning: applications of physics: 1/2 year).
7. The power of illustration: visual explanation and communication (Toning: Interdisciplinary Master's thesis in Physics and Communications).
8. Physics education: planning and execution of a course of education (Toning: 1/2 year).

9. Modeling of a windmill wing (Toning and Internal: 1/2 year).
10. Course of education on oscillating systems: synopsis of an educational experiment (Toning: 1/2 year).

JVM participated in two RUC projects: one successful, one not.

Project 1. Basic Studies project, natural science: Stellar astronomy.

This group, fulfilling the requirements of semester 2 (models, theories, experiments), set out to make measurements of the acceleration of the Universe's expansion. They selected as their advisor an experimental physicist whose specialties were condensed matter and biophysics. Thus, she could act as an informed guide, but not a specialist. Her guidance led, appropriately, to the students' quickly realizing that their chosen project was well beyond the scope of what was feasible. Through preliminary readings, and discussions among themselves and with the advisor and me they soon selected a more limited and feasible topic: stellar spectroscopy. This in itself is no easy task in habitually overcast Denmark. Nevertheless, the students designed and built a simple spectroscopic camera, took photographic spectra of stars, and analyzed the spectra to ascertain chemical compositions. They read widely on stellar structure and atomic spectroscopy. (One of the initial group members, who had not attended meetings, was given written warnings by the group, and eventually dropped out.) The group made several modifications of its project, based on the mid-semester evaluation—a process of several hours—conducted by the advisor and by another group and its advisor. Most of the modification was in reorganization of the material. The final evaluation/examination was conducted by the group's advisor and an advisor from yet another group. Each student presented a part of the project and received an individual evaluation. The group also received a joint evaluation. Finally, the students evaluated each other. The product was a monograph, entitled, with classic Danish irony, 'A Lovely Night in Hestekøb.' Hestekøb is a town in the often-foggy Copenhagen area. Finding an adequate, let alone lovely night for stellar spectra photography was no mean feat.

Project 2. Advanced Studies Physics Metaproject (1/2 year): Quantum nonlocality.

One of the presumptions at RUC is that no project should have a prerequisite, that everything needed will be learned in the project. This project proved that presumption wrong. The group set out to understand the debates between Bohr and Einstein regarding the meaning of quantum mechanics: Is its probabilistic description of nature fundamental, or is quantum mechanics an incomplete theory, waiting for a better, deterministic theory to replace it? The group's advisor was well-versed in the area, and provided them with considerable assistance in selecting readings. Both he and I lectured to the group on several aspects of quantum nonlocality. Unfortunately, while the students had had the equivalent of a first course in Modern Physics, none had any but the most rudimentary knowledge of quantum mechanics, and they were thus unable to grasp the essentials of the Bohr–Einstein debates. The advisor attempted to recast the problem in non-physics terms. Correlation without causation, a common approach to Bell's inequality, was described in terms of questionnaires administered to separate families by two sociologists who never showed each other their results. This novel approach, while intriguing to someone already conversant with the issues, was lost on the students. After much effort on the part of all of us, teachers and students, the group ultimately decided to change topics. This occurred late enough that they had to defer the project until the following semester.

Nevertheless, a version of the RUC model, but with prerequisites, could be established elsewhere. At LUC for example, project work could reside not, or not only, in the major or elective segment, but in the general education requirement. Physics majors might for example do project work to fulfill one of their social science requirements. They would be working in groups with students from other majors and they would learn by hands-on practice the modes of inquiry of a social science; this is not the usual outcome of a survey course in sociology for non-majors (if such exists). Similarly, non-science majors could do project work with physics students as part their science requirement, as an alternative to the traditional Physics for Poets.

This model, while very creative intellectually, may be hard to sell—it really obliges faculty to do something different with non-majors, well beyond a scaled-down version of the lecture courses for majors. But it also opens up the possibility of doing interdisciplinary project work at the junior and senior levels. Just as team-taught interdisciplinary courses provide creative alternatives to traditional course work, so can multidisciplinary projects. That being said, failures can be expected. Science and Society: Religion and Science, co-taught by JVM and a theology professor, was not a success. Our teaching styles were too different and the interaction between religion and science was not clear. The students in their post-course evaluations said that it felt like two different courses given at the same time.

The costs of project work are numerous. It is labor intensive. It is limited to groups of no more than ten; thus, a faculty member would have to supervise at least two groups to match the full-time-equivalent teaching load of a single modestly enrolled course, although perhaps not at the advanced level. Its coverage does not duplicate that of the traditional course. It requires a serious modification of traditional content and teaching practice. It demands training and supervision of faculty on an ongoing basis. But the realized and potential advantages are also numerous:

- Interactive engagement at a very high level.
- Enhancement of student performance.
- Building of student confidence.
- Improved retention in the major and at the university.
- Gender and ethnic equity.
- Cooperation instead of competition.
- Career preparation.
- The possibility of a profound interdisciplinarity.

10.4 Pedagogical theory

Three pioneers in pedagogical theory were Lev Vygotsky, Jean Piaget, and Jerome Bruner.

Lev Vygotsky (1896–1934) was a Soviet psychologist who is often named with Swiss psychologist Jean Piaget as founders of constructivism. Vygotsky is especially remembered for:

1. His distinction between one's *inner speech*, *private speech*, and *external speech*. Preschool children frequently talk out loud to themselves as they play and explore the environment. This private speech develops into inner speech, spoken within your head. Inner speech is not the interior aspect of external speech—it is a function in itself. It still remains speech, i.e. thought connected with words. But while in external speech thought is embodied in words, in inner speech words die as they bring forth thought. Inner speech is to a large extent thinking in pure meanings. It is a dynamic, shifting, unstable thing, fluttering between word and thought, the two more or less stable, more or less firmly delineated components of verbal thought [15]. External speech is what is primary in group projects. So is the transformation of inner speech to external speech—a non-trivial activity.
2. The Zone of Proximal Development (ZPD; figure 10.4) is 'the distance between the actual developmental level as determined by independent problem solving and the level of potential development as determined through problem solving under adult guidance or in collaboration with more capable peers' [16].

The ZPD is where new learning takes place, with help either from the teacher or from more capable peers. The latter is of prime importance in group projects.

Jean Piaget (1896–1980) was a Swiss biologist who is most famous for his four-stage model of children's developmental stages:

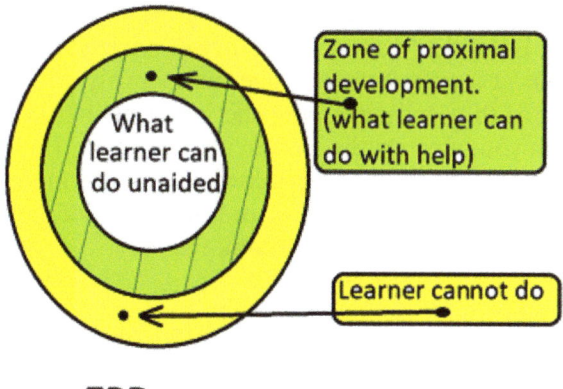

Figure 10.4. Zone of proximal development.

1. *Sensorimotor stage*: 0–2 years.
2. *Pre-operational stage*: 2–7 years.
3. *Concrete operational stage*: 7–11 years.
4. *Formal operational stage*: 11–16 years and onwards, with development of abstract reasoning. [17]

The year-spans given are Piaget's and were approximative. One could therefore suppose that the students we deal with in this book were at the formal operational stage. But later studies have shown that many 16-year-olds had not reached it and may never do it. And some students were stage 4 in some areas but not in others. Numbers differ, but apparently 10%–20% of our student body is not formal operational, i.e. ready to:

- Systematically solve a problem in a logical and methodical way.
- Think about abstract and hypothetical ideas.
- Ponder 'what-if' type situations and questions and think about multiple solutions or possible outcomes.

These are the three characteristics of the formal operational stage.

It is crucial in group projects that one can contribute in a logical and methodical way, can argue about abstract and hypothetical ideas, and can suggest 'what-if' questions.

There may well be a connection between operational stages and science anxiety [17]. Can a student who is formal operational under most circumstances drop back to concrete operational during, for example, the stress of an exam? Anecdotally, we have seen this in our own classes.

Jerome Bruner (1915–2016) was an American psychologist, active over a very long span of years. Among his many contributions was the concept of *scaffolding*. It was a further-development of Vygotsky's idea of the zone of proximal development. When students work in their ZPD one should support them by building up helping systems. Bruner argued that one should narrow the number of degrees of freedom to make their journey through ZPD easier. This is contrary to the tenor of this book. Generally, we argue that assignments, group projects, and labs should have many degrees of freedom. But if the freedom is a source of anxiety, there are good reasons to follow Bruner. Or at least to follow Bruner temporarily.

When you argue with a colleague why she is not giving the students numbers (degrees) of freedom you often get the answer: 'No, not now. First, they will have to learn the tricks of the trade. Later I will slowly introduce more freedom.' This a dangerous way of thinking. If you teach them how to work following a recipe, you train parrots who will not easily begin to think laterally.

References

[1] Hake R 1998 Interactive engagement vs traditional methods: a six-thousand student survey of mechanics test data for introductory physics courses *Am. J. Phys.* **66** 64–74

[2] Solomon J 1994 Constructivism and quality in science education *Naturfagenes Pædagogik* ed A C Paulsen (Gilleleje: Roskilde University Press) pp 17–29

[3] Gautreau R and Novemsky L 1997 Concepts first—a small group approach to physics learning *Am. J. Phys.* **65** 418–28
[4] Heller P, Keith R and Anderson S 1992 Teaching problem solving through cooperative grouping. Part 1: group versus individual problem solving *Am. J. Phys.* **60** 627–36
Heller P, Keith R and Anderson S 1992 Teaching problem solving through cooperative grouping. Part 2: designing and structuring groups *Am. J. Phys.* **60** 637–44
[5] Mazur E 1997 *Peer Instruction: A User's Manual* (Upper Saddle River, NJ: Prentice Hall)
[6] Meltzer D and Manivannan K 1996 Promoting interactivity in physics lecture classes *Phys. Teach.* **34** 72–6
[7] Michaelsen L, Watson W, Cragin J and Fink L 1982 Team learning: a potential solution to the problems of large classes *Exchange: Organ. Behav. Teach. J.* **7** 13–22
[8] Treisman U 1992 Studying students studying calculus: a look at the lives of minority mathematics students in college *Coll. Math. J.* **23** 362–72
[9] Beyer K 1992 Project organized university studies in science: gender, metacognition and quality of learning *Contributions to the European GASAT Conf., Eindhoven* pp 363–72
[10] Giese P 1997 Learning through project work ed J Mallow *Physics Teaching in Scandinavia: A Summary Monograph AAPT Committee on International Education* (College Park, MD: American Association of Physics Teachers) pp 28–34
[11] Mallow J 1993 *Problem-Oriented Group Project Work at Roskilde University IMFUFA Text 336* (Roskilde: Roskilde University Press)
[12] Steele C and Aronson J 1995 Stereotype threat and the intellectual test performance of African Americans *J. Pers. Soc. Psychol.* **69** 797–811
[13] Osborne J 2007 Linking stereotype threat and anxiety *Educ. Psychol.* **27** 135–54
[14] Mallow J 2001 Student group project work: a pioneering experiment in interactive engagement *J. Sci. Educ. Technol.* **10** 105–13 and references therein
[15] Vygotsky L 1962 *Thought and Language* (Cambridge, MA: MIT Press)
[16] Vygotsky L 1978 *Mind in Society: The Development of Higher Psychological Processes* (Cambridge, MA: Harvard University Press)
[17] Hsueh Y 2005 The lost and found experience: Piaget rediscovered *Constructivist* **16** 1–11
[18] Wood D J, Bruner J S and Ross G 1976 The role of tutoring in problem solving *J. Child Psychol. Psychiatry* **17** 89–100
[19] Mallow J V 1986 *Science Anxiety and How to Overcome It* (Clearwater, FL: H & H Publishing)

Chapter 11

Pedagogies for different student populations: they're not dumb, they're different

They're not Dumb, They're Different is the title of a book by Sheila Tobias [1]. The subtitle is *Stalking the Second Tier*, by which is meant students who have an interest in science, although they will be studying something else. The book comprises detailed interviews with such students; actually, graduate students with non-science degrees. Each enrolled as an auditor in an undergraduate science course and reflected on their experiences therein. By and large these were not positive, and in one case a potential science student chose a different path after experiencing a science class. In any case, that was not the goal of the study, which was to see what might be off-putting to non-science students and how that might be changed. Among the answers were the goals of the class, the nature of the pedagogy, and the classroom atmosphere. To paraphrase the students, 'I want to understand the big questions about physics and the world, but the class is focused on problem-solving'; 'The format is entirely lecture'; 'There is little interaction between students'. All of this is painfully true in many courses, although much of that has changed since the Tobias book with, *inter alia*, constructivist pedagogy (chapter 4), group work (chapter 10), and courses dealing precisely with the relationship of physics to society (chapter 12). In this chapter, we shall consider the various student populations who enroll in introductory physics classes, either because it is a requirement, or because it is a field of interest. We shall describe the range of approaches, what they have in common and how they differ[1]. We shall consider students' attitudes and anxieties. We shall describe both extant and desired pedagogies.

We begin with observations relevant to all students, from physics students to liberal arts students taking a physics course. In many universities those teaching even introductory physics courses are primarily dedicated to (and have been hired for)

---
[1] 'Compare and contrast', to coin a phrase.

research[2]. Teaching is seen as an odious chore[3]. Questions and answers are delegated to graduate assistants. So are grading of homework and exams. These generally contain only numerical problems, rarely any essay questions. There are even online grading programs that interact with the student in the following way: the student tries to solve a homework problem. If they get stuck, they can ask the program for help. It will give them enough information to go further—but *will award them a demerit for asking*. At the end of the semester, the accumulated demerits are factored in, and voila! Neither instructor nor graduate assistant need waste time grading. We can hardly think of a better mechanism for producing science anxiety.

Another glaring lacuna is the lack of pedagogy instruction for prospective college and university teachers. Why should it be that primary and secondary school teachers should be required to take education courses, while professors are expected to know how to do it by instinct[4]? Some do, some don't. (Results of the latter are legion.) Short of introducing pedagogy courses for professors[5], mentoring is a solution. An experienced faculty member is designated to visit the mentee's classroom, perhaps every third session, and then discuss what has been done well and what can be improved.

11.1 Evaluations

A prominent source of anxiety is the dual role of the instructor: teacher and evaluator. In fact, one could say that the roles are in actual conflict. How can a student ask questions of the teacher when they fear that it can make a negative impression? Whether or not this is objectively true is irrelevant to the anxiety.

Evaluations are of four types: teacher–student, teacher–teacher, student–teacher, and student–student.

11.1.1 Teacher–student classroom evaluation

Methods for evaluating student performance vary widely, both between and within IE and traditional courses. We have taught both IE courses, such as group projects, and courses with a mixture of traditional and IE. For IE courses such as group projects, evaluation usually comes at the end with a presentation, but the teacher is (presumably) interacting with the students throughout the project. For more traditional courses, evaluation is done by examinations. These are a source of anxiety for many students. In many European countries this consists of a single test at the end of the course. That places enormous pressure on the student, and is not a good measure of performance. What if the student just has a bad day? The result can be failure of the entire course. The typical US model is different, and less anxiety-producing.

---

[2] In our acerbic moments, we describe them as 'research institutes masquerading as universities'.
[3] A sterling counterexample was JVM's teacher at Columbia University, Nobel Laureate Polykarp Kusch, who won an award for his teaching of introductory physics to students at Barnard College, Columbia's (then) sister school.
[4] Both of us have studied and taught pedagogy.
[5] Although why shouldn't we?

It consists of several tests plus a final exam, at minimum a midterm and a final. The grade is normally a weighted average of the two. More common is the model of two or more well-spaced tests, plus a final. Even more, in many classes, the lowest grade on one of the tests is automatically dropped, and the grade becomes the weighted average of the others and the final. This builds in the 'bad day' correction, or family matters such as illness or death of a relative[6].

The teacher can do various things to give students more confidence in the classroom. We referred in chapter 3 to the American Association of Physics Teachers' student confidence workshop [2]. The teachers were asked to assess their relative focus on content versus relationship (see appendix A). The desired outcome of the workshop was to have them strive for balance between the two, to cover the material while maintaining a classroom atmosphere that minimized anxiety and maximized confidence. We consider one example: the wait time between the teacher's posing a question and the students' answering it. If the teacher poses the question to the whole class, then the quicker (in either sense of the word) students will shoot their hands up and the teacher will call on one of them. This procedure disadvantages the other students, not allowing them enough time to think the question through, and thus lowering their confidence. The same holds for questioning single students. The teacher's impatience will be manifest in a short wait time[7]. Of the various ways of counteracting this phenomenon, we have developed one which we think works well. We call it the modified Socratic method. Don't ask for hands. Choose a student (preferably one who does not usually shoot their hand up) and ask her or him the question. Then wait. Don't rush. Either the student will answer in a reasonable time, or cannot. If the latter, follow up with a question whose answer led up to it. If the student still cannot answer, go back further. In our experience, almost all students will reach a point, perhaps several questions back, when they can answer. Then bring them forward. For example, going back:

> Conservation of energy ⟹
> Definitions and examples of potential and kinetic energy ⟹
> Definition of energy.

There are two ostensible difficulties with this method. First is the time consumed. Second is the possibility of provoking the student's anxiety. For the first, the time used is a review of the subject for the other students, some of whom are unsure of the various answers. For the second, it is the teacher's obligation to make clear that this is the case, and to assure the student—and all the others—that the procedure will be

---

[6] When JVM's wife was teaching foreign language, this model was not in effect. Excuses had to be produced for missed exams. One student claimed to have lost three grandmothers, each on a test day. She rationed him to only one.

[7] This has also been shown to be gender biased: shorter wait times for females before moving on.

used randomly on all the students[8]. Finally, the teacher should praise the student as each of the questions is answered.

To summarize:

- The combined roles of teacher and evaluator are a source of anxiety.
- A balance of relationship and content can mitigate this anxiety.
- This can be accomplished by interactive engagement.

11.1.2 Teacher–student group project evaluation

The evaluation takes the form of a final presentation, most efficaciously accomplished in the presence of other instructors and students, all of whom can pose questions to the presenters. The instructor can take note of these, as well as any criticisms. Based on these observations and his or her own assessment, a group grade can be determined.

In addition, the instructor needs to make two other determinations.

- Was the project of appropriate difficulty? Since the instructor was (or should have been) part of the proposal process, as well as its advisor throughout the process, she or he should be able to assess this. Somewhere in the process, the instructor should have been able to determine the answer. If the answer is yes, then the evaluation of the final presentation can be compared with the proposed project. If no, then in the course of the project, the instructor, in discussion with the students, should have been able to modify it.
- In addition to the group evaluation, there must be an individual evaluation: the relative contribution of each group member. Who did what? Each member of the group should be required to present some component of the work. The final grade should be a combination of the group and the individual grades.

11.1.3 Teacher–teacher evaluation

There is little we can say here that is not already known and practiced. The list includes:

- Class visits by fellow faculty.
- Annual review, including self-evaluation and discussion with chair and possibly dean.
- Midterm review, third year, by faculty and chair.

---

[8] It is a well-known fact that irrespective of the teacher's gender he/she will automatically ask many more questions to male students than to female. One has to use some sort of systematics so that male/female, black/white Muslim/Jewish/Christian, etc, students are asked the same percentage of questions as the size of their group requires. HK once checked the statistics in his own teaching, though he was sure it was unnecessary—it wasn't. He began asking questions systematically according to where they were sitting. First students in the back row etc. This has of course to be done diligently.

- Review for promotion and tenure, sixth year, by faculty and chair with internal and external references.
- Review for subsequent promotion, by faculty and chair with internal and external references.

11.1.4 Student–teacher evaluation

These generally take the form of anonymous questionnaires, 'teacher–course evaluations' (TCEs). They include both numerical value questions and brief comments. For the most part, they are similar across departments and even universities, although to our knowledge there is no standard form. They play an important role in the annual review and assessment. While useful, they may be used inappropriately. (We know of a dean who determined raises based on the third digit of the TCEs[9].) TCE averages are of limited value, except when the responses are strongly positive or negative. A general rule might be, look for the outliers. Strongly negative ratings and comments about an instructor point to the need to monitor and help improve that instructor's teaching skills. This can include content delivery and relationship (see [2] and appendix A). One or several student evaluations and comments that are strongly negative compared to that of the class as a whole should be viewed with suspicion. TCEs can be abused, especially in small classes. 'Revenge TCEs' are not uncommon. In one case at Loyola, several students colluded to give their instructor bad ratings.

11.1.5 Student–student evaluation

These are *ipso facto* restricted to projects. At RUC, for example, students provide the same individual evaluations of their group members as does their instructor. These are heavily based on whether individual members pulled their own weight. In some cases, the group fires a member for not doing so.

11.2 Teaching physics students

In chapter 1 we mentioned Hake's seminal meta-study comparing gains in understanding mechanics in traditional versus interactive engagement courses [3]. Here we address it in more detail. Hake obtained data from 62 introductory physics courses in high schools, colleges (including community colleges), and universities comprising a total of 56 542 students, that had introduced various versions of IE, and had compared results with traditional courses. We shall consider only the 32 universities for which 'introductory physics' means, in part or whole, courses for physics and/or engineering students. Among Hake's list of IE methods implemented by various universities were [4]:

- Collaborative peer instruction.
- Employment of undergraduate students to augment the instructional staff.
- Context-rich problems.

---

[9] Not in one of our institutions.

- Interactive simulations with worksheets.
- Cooperative group problem-solving.
- An inquiry approach with qualitative questions and experiment problems in the labs.
- A course web page.
- Computer communication between and among students and instructors.
- Team teaching.

In most cases, IE courses produced significantly larger knowledge gains than traditional courses. Of the ones that showed only low or moderate gain, Hake observed that this may have been the difficulty in implementation of IE. Class sizes in some of these exceeded 100 and sometimes 200. Attempts to utilize IE appear to have been overwhelmed by sheer class size.

Time has passed since Hake's study, and more IE methods have been introduced in many courses, but introductory physics class sizes have not been reduced in many, perhaps most, institutions, at least in the US. IE methods are thus undercut.

In chapter 4 we also described the study of Gautreau and Novemsky [5], in which, despite the clear evidence that IE was superior, independent of teacher, they all went back to lecturing. Inertia? Fear of leaving one's comfort zone? The daunting task of full course revision? These obstacles need to be overcome. The first two are psychological, which are the most difficult, although some discussion in faculty meetings might be of use. The third can be dealt with by designating time (by course reduction, not overload) to help teachers carry out the task. To be clear, we are not proposing the disappearance of the lecture format. Courses can be a mixture of traditional and IE. In the simplest case, the lecture should be interspersed with class discussion.

But why not teach our physics students the way that we were taught—the traditional chalk and talk by the sage on the stage? It worked for us. If it ain't broke, don't fix it. However, Hake's study of the 32 universities' courses showed that even for physics students, IE produces greater gains in understanding than does the traditional method, and in the same amount of time. As we have already noted, Tobias's study showed that students who might want to study physics or science in general can be driven away by some of the features of traditional teaching.

The argument is frequently made that even if IE produces better understanding of a topic such as mechanics, the teacher still has to get through the curriculum, and only lecturing can do this. But in the Hake meta-study, there was no difference between traditional and IE courses in time or coverage allotted to mechanics, so the same total curriculum can be covered. While there is no reason to believe that this would not hold true for the other topics, it should be tested.

11.3 Teaching pre-health students

The teaching of physics to pre-health students, i.e. those applying to schools of medicine, dentistry, podiatry, and others, presents special problems. The first is 'teaching to the test'. The Medical College Admission Test (MCAT) is a computer-based standardized examination for prospective medical students in 20 countries[10] [6]. Questions are multiple choice. The physics section contains about 15 questions, to be covered in 25 min. They cover everything from classical mechanics through atomic and nuclear physics. This is, to say the least, daunting.

Compounding the problem is class size. In the US, pre-health students are overwhelmingly majoring in biology. Only if several sections are provided can these numbers be reduced to tractability. At Loyola, no physics class comprises more than 60, still rather large. Group projects are probably infeasible. But interactive engagement is not. A combination of lecture and in-class exercises can work, as it does for the other student cohorts.

In addition, for a good number, perhaps the majority, of the pre-health students, physics is just a course to 'get through', cramming in as much material into as short a time as possible. Many professors agree. Pure lecture courses, although shown to be less effective that IE, still constitute the norm. But as we noted above, as long as there are classes with hundreds of students, IE does not work.

Last but not least is the title of this book. Many of the pre-health students are science-anxious, especially about physics. The techniques discussed here [2, 3, 5] and elsewhere [7] should provide teachers with sufficient tools to alleviate anxiety among pre-health students, much as they do for other cohorts. They might also change attitudes so that physics is more than just a gauntlet to be run.

11.4 Teaching liberal arts students

In this group we include humanities and social science students. They are taking physics either because it is required by the university or it is an elective in which they are simply interested. In any case, we do not wish to make any distinction. The only requirement is that a course be designed for them. The list of issues is the same as for the other cohorts, briefly:

- Anxiety reduction.
- attitude assessment and modification.
- Interactive engagement.
- Group work.

Each of these should be addressed, sometimes in the same way as for the other cohorts, sometimes differently. In the case of anxiety reduction [2] and [7] provide techniques appropriate for all of the cohorts. That being said, the attendees of the Loyola Science Anxiety Clinic were typically science students [8]; the numbers of non-science students were suspiciously low. We believe that they took the avoidance option. If any science was required, they had the choice of taking courses other than physics. To this problem, the only answers are university-wide or nation-wide campaigns to explain that learning physics is essential to society and is interesting to boot. It should not be optional. Here we discuss students who have opted in. For liberal arts students the goal is not coverage, but exemplarity. We mentioned this in chapter 10 in the context of group projects, but its application extends beyond that. In particular, the one-semester course

---

[10] The actual numbers listed on the MCAT and Wikipedia sites vary from 17 to 21.

can limit itself to one area. The underlying philosophy is that one area teaches students not only some content, but is an exemplar of how physics works—our ultimate goal. The one-semester course that JVM taught at Loyola covered mechanics. The prerequisites were high school algebra and geometry.

Topics covered were:

- Describing motion.
- Falling objects and projectile motion.
- Newton's laws: explaining motion.
- Circular motion, the planets, and gravity.
- Energy and oscillations.
- Momentum and impulse.
- Relativity.

Note the inclusion of relativity, special and even a bit of general. The only mathematics needed were simple algebra (e.g. no simultaneous equations) and very simple geometry: the Pythagorean theorem.

The class was a combination of lecture and interactive engagement, the latter including class discussion, the modified Socratic method, and small in-class group exercises. For example, for centripetal force, the instructor and students worked a few problems together. Then the instructor asked the students to break into groups and explain how clothes washers and dryers work, with numerical examples. In lieu of a laboratory, students were asked to design and carry out simple home experiments. Students were also expected to learn from interactions between them and the teacher, and among themselves.

There are two special components to the course each Friday. The first is general question–answer time. For a few minutes at the beginning of the hour, the students pose any questions you may have about science in general, as well as its interaction with society. This allows for the coverage of issues that fall outside the syllabus. The second is the Critical Incident Inventory, shown below and in appendix A. This is a brief anonymous questionnaire that provides feedback to the instructor:

1. What was your most engaged moment?
2. What was your most distanced moment?
3. What was the most affirming or helpful action?
4. What was the most puzzling action?
5. What surprised you the most?

Below is a list of sample questions that the students should be able to answer by the end of the course.

11.4.1 Scientific methodology

- Among the types of statements that make up physics (or any science) are beliefs, definitions, empirical descriptions of what we observe, and laws or theories. For each of the following, state which type of statement it is.

a. The force of gravity between any two masses varies inversely as the square of the distance between them.
  b. Acceleration is the rate of change of velocity over time.
  c. The Universe is orderly.
  d. Planets move around the Sun.
- Describe one way which physics and another field of study, such as your major, are similar and one way in which they differ.
- Newtonian mechanics works for macroscopic objects, but fails for microscopic objects, where quantum mechanics is correct. Does quantum mechanics fail for macroscopic objects?

11.4.2 Newtonian physics

- A 1.30 kg model boat crossing a stream is subject to three horizontal forces: a force of 12.0 N east due to its motor, a force of 7.00 N south due to the current, and a force of 2.00 N north due to the wind.
  a. What is the magnitude and (approximate) direction of the boat's acceleration?
  b. If the boat starts from rest, how far has it gone after 4.50 s?
- What would a woman who weighs 120 lbs on Earth weigh on Neptune, whose mass is 17.24 Earth-masses, and whose radius is 3.81 Earth-radii?
- A planet takes 81 years to orbit the Sun. What is its average distance from the Sun?
- A student with her feet on the floor pulls on a rope attached to a professor on a skateboard, twirling him in a horizontal circle of radius 1.50 m.
  a. What is the professor's speed?
  b. It is a common misconception that there is no net force on an object in circular motion. What could the student do to disprove this, and what would be the consequences to the professor?

11.4.3 Relativity

- Galaxy A is receding from the Milky Way (our Galaxy) at 0.75c. Galaxy B is receding from the Milky Way in the opposite direction at 0.5c. What does someone in Galaxy B measure the speed of Galaxy A to be?
- You are in a windowless room.
  a. If the room is an elevator on Earth, ascending with an acceleration of 4.9 m s^{-2}, how much heavier do you feel than normal? Give your answer as a fraction or a percent.
  b. If you felt the same as in part a, what else, other than an elevator on Earth, might the room be, and what would it be doing?

11.4.4 Quantitative analytical skills

The graph describes the motion of a vehicle over 10 s.

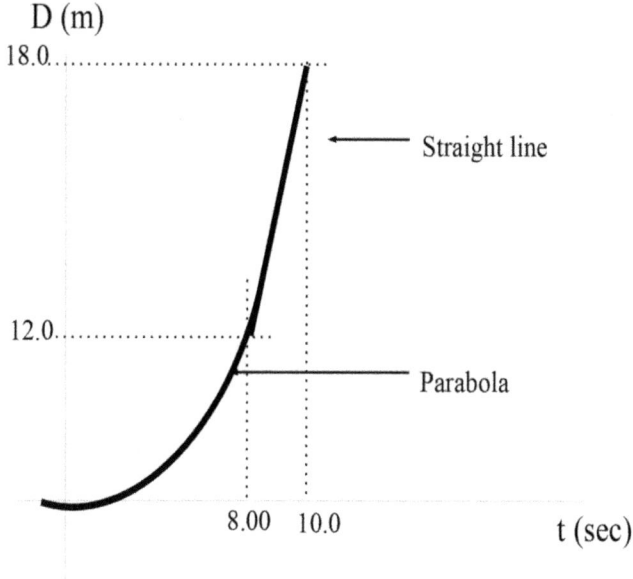

a. For each of the two regions, 0–8 s and 8–10 s, state whether the vehicle is sitting at rest, moving at constant speed, accelerating, or decelerating.
b. Graph qualitatively the speed versus time.
c. What is the car's average speed over the entire 10.0 s?

Another course (also taught by JVM) was 'Energy for a Sustainable Future'. The course was described as follows: 'The physics of energy, the technology of its production and use, and the environmental and socio-political implications of energy resources and distribution will be the focus'. It represented exemplarity, project work, and physics and society. The learning objectives were the following:

- To understand the qualitative nature of physics by focusing on a subset of topics in physics, in this case, energy.
- To understand the interaction between theory and experiment.
- To understand the quantitative nature of physics by solving problems using algebra and geometry.
- To understand the process of learning physics by engaging in group discussion.
- To understand the values nature of physics by posing questions about physics and society, especially as regards energy production, distribution, and usage.
- To distinguish between scientific and non-scientific problems.

The pedagogies are the same: interactive engagement, modified Socratic dialogue, group work, and feedback to the instructor, including the Critical Incident Inventory. The topics covered were:
- Energy and society.

- Energy principles.
- Electric energy.
- Global warming and ozone depletion.
- Nuclear physics principles.
- Nuclear fission.
- Fission breeder reactors.
- Nuclear fusion.

Among the projects were:
- Study of geothermal energy: a research paper.
- Study of wind energy: measurement of wind velocities on the shore of Lake Michigan.
- Study of nuclear fission: features of a local nuclear reactor.
- Study of solar energy: measurement of solar radiation at various locations.

References

[1] Tobias S 1990 *They're not Dumb, They're Different* (Tucson, AZ: Research Corporation)
[2] Fuller R, Agruso S, Mallow J V, Nichols D, Sapp R, Strassenburg A and Allen G 1985 *Developing Student Confidence in Physics* (College Park, MD: AAPT) workshop manual
[3] Hake R 1998 Interactive-engagement versus traditional methods: a six-thousand-student survey of mechanics test data for introductory physics courses *Am. J. Phys.* **66** 64
[4] Hake R 1998 Interactive-engagement versus traditional methods: a six-thousand-student survey of mechanics test data for introductory physics courses *Am. J. Phys.* **66** 64 (references)
[5] Gautreau R and Novemsky L 1997 Concepts first—a small group approach to physics learning *Am. J. Phys.* **65** 418–29
[6] https://students-residents.aamc.org/taking-mcat-exam/taking-mcat-exam
[7] Kastrup H and Mallow J 2016 *Student Attitudes, Student Anxieties and How to Address Them: A Handbook for Science Teachers* (San Rafael, CA/Bristol: Morgan & Claypool/Institute of Physics)
[8] Mallow J V 1986 *Science Anxiety: Fear of Science and How to Overcome It* (Clearwater, FL: H & H Publishing)

IOP Publishing

Fear of Physics
And how to help students overcome it
Jeffry V Mallow and Helge Kastrup

Chapter 12

Physics and society: close encounters

In September 1939 the Second World War started with the German invasion of Poland. The Danish doctor Ole Chievitz raised the question of a major gas attack that Danish hospitals were not prepared for. He suggested that one should produce 6000 devices to be distributed to all hospitals, enabling them to give oxygen treatment to victims. But no factories in Denmark were able to produce them at short notice. Professor Chievitz asked his close friend Niels Bohr for advice. Bohr promised to help. Immediately all the employees at the Niels Bohr Institute[1], theoreticians as well as experimentalists, were involved in producing a simple version of the device, and within a week they had the 6000 needed.

This is one anecdote among many about the social engagement of physicists. In contrast there is a plethora of anecdotes about absent-minded physics professors and teachers with low social skills. For instance: A very absent-minded physics professor entered a crowded bus, with no available seats. Suddenly a little girl rose from her seat and offered it to the professor. He was astonished and said to her: 'You are a very good girl, what's your name?' 'My name is Eve, daddy.'

The two stories are not each other's antithesis. Absent-minded teachers exist in all subjects and might very well be strongly socially engaged nevertheless. And some teachers, including physicists, do their jobs excellently, even with low social contact and engagement. But they ought perhaps not to teach. In this chapter we will consider three very different aspects of physics and social engagement.

12.1 Close encounters of the first kind: the legendary cold inhuman physics lab

Just as a doctor's surgery can be scary if you are afraid of scalpels and syringes, so can the physics lab be, with its high voltage outlets, stroboscopic lights, and lasers

---

[1] Everybody called it Niels Bohr's Institute from day one, but it eventually got its name officially in 1965 after Bohr's death.

Figure 12.1. Physics lab in Copenhagen at a Catholic girls' school in the beginning of the 1920s.

(see chapter 9). Indeed, labs were traditionally sterile looking places. That was (and in many cases still is) true of the classroom as well. Figure 12.1 shows the lab of the physics class that the mother of HK attended in the beginning of the 1920s. It does not look very welcoming to the female adolescents being taught there.

Fortunately, few labs today look so uninviting. They are decorated with pictures of Richard P Feynman, Niels Bohr, Albert Einstein (sometimes sticking out his tongue), Tycho Brahe's wall quadrant for determining time by the altitudes of stars, and more and more often pictures of female physicists and astronomers such as Emmy Noether, Marie Curie, Lise Meitner, Maria Goeppert Mayer, Chien-Shung Wu, and Vera Rubin. We have easy access to hundreds of posters of galaxies, planetary nebulae, gas nebulae, high energy collisions, and fusion reactors. But more important is the spirit of togetherness on a journey of learning, understanding, and exploring. A teacher should be engaged with her students. Beyond the class and lab, this means being a whole person. This does not mean that the teacher is their comrade. But they are, or could be, a role model. It is particularly important that a growing number of females are present in our ranks (chapter 6). We are pushing at an increasingly open door. The teacher can help students of common gender or ethnicity overcome stereotype threat (see chapter 11). 'If this Black woman/Muslim/Inuit teacher can, then so can I.'[2]

---

[2] One of us (HK) taught for 19 years at different teacher training colleges. Two of those had about one-third Muslim students, many of whom were girls in (beautiful) headscarves. When we had open-house arrangements trying to recruit new students, it was obvious that when the young Muslim girls saw many others like themselves, they chose us.

Another way of being engaged with the students is to arrange extracurricular activities. HK has led science tours to London and Oxford, visiting places like the Royal Institution, the JET Laboratories, the Natural Science Museum, the London Science Museum, the Greenwich Observatory, and others. In particular, it is always good to make excursions to scientific facilities. There are many possibilities, such as power stations, science museums with hand-on experiments, lighthouses, and chemical plants. Some can be part of group projects (chapter 10). Since many technical industries and universities are well aware of the scarcity of upcoming employees, it is quite common for them to provide programs in which students can experience at first hand what is being done.

To summarize, it is the obligation of physicists to make both the physical environment and the interaction with students inviting, in both senses of the word: comfortable for those already there and welcoming to those who are not yet there, but well might come.

12.2 Close encounters of the second kind: physics as the confluence of the legendary nerds

There are some students who have no need of the invitations discussed above. They wanted to be there from the start—and with a vengeance. Some of them are what society calls 'nerds'. To us, being called nerds is not objectively negative, just insulting. We don't subscribe to the Wikipedia definition, 'A nerd is a person seen as overly intellectual, obsessive, introverted or lacking social skills. Such a person may spend inordinate amounts of time on unpopular, little known, or non-mainstream activities.'[3] To us such a person is an intellectual who spends time on a subject needing a lot of effort. Physics, indeed all the sciences, are such subjects. Our labs are free havens for such nerds. It is good that we have a treasured place for those who will probably turn into physicists, but it is a double-edged sword. Society needs many more students learning physics for other lines of studies such as medicine, midwifery, nursing, engineering, and archeology. Equally important is producing educated citizens capable of understanding discussions and decisions on energy politics, nuclear power, renewable sources, pollution, space exploration, nuclear weaponry, CO_2 emissions, as well as the IT-revolution (chapter 13)[4]. Therefore, while we welcome the physics nerds into our realm, it must obviously be done in a way so as not to scare the other students away[5]. Otherwise we will be seen as what Fred Hoyle called 'high priests of a not very popular religion'. However, we are not speaking of a dichotomy between loving or loathing physics. The dangerous dichotomy lies between intellectuals whose breadth of understanding includes science, and, if you will pardon our rancor, pseudo-intellectuals with no knowledge,

---

[3] Some people spend inordinate amounts of time on popular, well-known, mainstream activities such as reality TV shows.
[4] For our fellow senior citizens, we recommend that any IT purchase be accessorized with a grandchild.
[5] This is a general rule: whenever you make an effort to not scare a certain group—females, weak math students, non-nerds, specific ethnicities—you will help everybody.

understanding, or appreciation of the sciences. (The radical constructivists of chapter 4 fall squarely into this category.) There is a well-known exchange between an American film critic and a former editor of *Scientific American*, in which the critic proudly announces that she doesn't know any science. The editor replies, 'But shouldn't an educated person have some knowledge of all fields?' To which the critic retorts, 'Oh, a Renaissance hack!'

Nerds and non-nerds of the world unite. We have nothing to lose but our ignorance.

12.3 Close encounters of the third kind: physics for society

Chapter 2 has described the various negative attitudes that students have absorbed towards physics and its practitioners. In this section we turn instead to the ways that physics can contribute, and is contributing, to society. In addition to the well-known technologies such as CAT and PET scans, MRIs, or radiation therapies, we shall address several other contributions of physics to society.

12.3.1 Provision of correct information

Involvement in discussion forums, political work, or NGO activities can only be of value when grounded in physics. Its sound knowledge and methodology can help people evaluate the quality of different sources of information. Unfortunately it is very easy to find scientifically inaccurate news, opinions, and social media discussions of physics. Here are some examples of myths and their debunking:

1. CO_2 levels in the atmosphere and climate change https://www.wwf.org.uk/updates/10-myths-about-climate-change
2. Wind, hydroelectric, and solar power https://e360.yale.edu/features/three-myths-about-renewable-energy-and-the-grid-debunked
3. Nuclear energy https://www.anl.gov/article/10-myths-about-nuclear-energy
4. Melting of the Arctic and Antarctic ice shelves and water levels of the oceans https://climate.nasa.gov/news/3122/five-facts-to-help-you-understand-sea-ice/

12.3.2 Interaction with communities

Here are two examples:
1. Physics World https://physicsworld.com/a/scientists-collaborate-with-communities/. Its activities involve partnerships with local communities on issues such as flood prevention, pollution cleanup, and adaptation to climate change.
2. SENCER

SENCER is the acronym for Science Engagement for New Civic Engagements and Responsibilities. A full descriptions of its activities may be found at sencer.net. Its mission statement description says, 'SENCER courses and programs strengthen student learning and interest in science, technology, engineering, and mathematics (STEM) by connecting course topics to issues of critical local, national, and global importance.' An external survey (10 000 participants) of the effect of SENCER-designed courses [1] it was found that:

- Students gained most in the areas of *science literacy*, followed by general course skills.
- *Women* gained more than men and *non-science majors* gained more than science majors on many of the items and composite variables (a fact that the evaluators note 'is encouraging evidence given that females and non-science majors have traditionally been underserved or overlooked in many university science programs').
- The patterns of gains were in line with efforts by SENCER to encourage awareness of the *link between civic issues and scientific content*.
- Roughly a fifth of students who had never *engaged in civic activities* said they were more likely to participate in these activities after a SENCER course completion.
- Ten percent of students who on the pre-survey were not interested in *taking additional science or mathematics courses* reported on the post-survey that they were very or extremely interested in doing so.
- Similarly, slightly more than six percent now say they would like to consider *exploring career opportunities in science* and nearly five percent are now 'interested in teaching science'.

12.3.3 Research addressing societal issues

Cornell University provides a real-time list of research papers addressing societal problems both directly and/or using physics methodology [2]. Recent examples of each are respectively 'Increasing and diverging greenhouse gas emissions of urban wastewater treatment in China' [3] and 'Assessing the influence of French vaccine critics during the two first years of the COVID-19 pandemic' [4]. The full list comprises over 22 000 items. There are over 200 for January and February 2022 alone.

12.3.4 Individual initiatives

'Energy for A Sustainable Future' described in chapter 11, was a course designed by JVM at Loyola, for non-science majors. The physics of energy, the technology of its production and use, and the environmental and socio-political implications of energy resources and distribution were emphasized. It included group projects, such as:

    a. Acquisition and analysis of energy use per capita versus GDP per capita.
    b. Measurement of indoor–outdoor temperature loss through a window.
    c. Nuclear power plant visit and analysis.
    d. Measurement of wind velocity on the shore of Lake Michigan[6].

---

[6] Not as high as expected, demonstrating once again that 'Windy City' refers to the logorrhea of its politicians.

12.4 Conclusion

There is no clear boundary between 'basic physics' and 'physics for society'. There is no way to tell in advance whether a pure research discovery will have consequences for society. Magnetic resonance imaging (MRI) technology did not originate in applied physics, but in a fundamental discovery of an effect called 'nuclear magnetic resonance' (NMR). On the other hand, climate change research was not simply a matter of curiosity, but a matter of its critical effect on the planet. The International Union of Pure and Applied Physics (IUPAP) adopted the following statement on the importance of physics to society [5]:

- Physics is an exciting intellectual adventure that inspires young people and expands the frontiers of our knowledge about Nature.
- Physics generates fundamental knowledge needed for the future technological advances that will continue to drive the economic engines of the world.
- Physics contributes to the technological infrastructure and provides trained personnel needed to take advantage of scientific advances and discoveries.
- Physics is an important element in the education of chemists, engineers, and computer scientists as well as practitioners of the other physical and biomedical sciences.
- Physics extends and enhances our understanding of other disciplines, such as the Earth, agricultural, chemical, biological, and environmental sciences, plus astronomy and cosmology—subjects of substantial importance to all peoples of the world.
- Physics improves our quality of life by providing the basic understanding necessary for developing new instrumentation and techniques for medical applications, such as computer tomography, magnetic resonance imaging, positron emission tomography, ultrasonic imaging, and laser surgery.

What do we *not* do physics for? JVM has a T-shirt (mentioned earlier) that says, 'I BECAME A PHYSICIST FOR THE MONEY AND THE FAME'.

References

[1] Carroll S SALG results show SENCER faculty achieve in raising higher-order learning gains *SENCER Research and Results* http://sencer.net/sencer-results/#Gains
[2] Search results for physics and society *arXiv* https://arxiv.org/search/?query=physics+and+society&searchtype=all&source=header
[3] Huang Y, Shuming L, Meng F and Smith K 2022 Increasing and diverging greenhouse gas emissions of urban wastewater treatment in China arxiv: 2202.11511
[4] Faccin M, Gargiolo F, Atlani-Dualt L and Ward J 2022 Assessing the influence of French vaccine critics during the two first years of the COVID-19 pandemic *PLoS ONE* **17** e0271157
[5] IUPAP 1999 23rd General Assembly of IUPAP https://archive2.iupap.org/general-assembly/23rd-general-assembly/

IOP Publishing

Fear of Physics
And how to help students overcome it
Jeffry V Mallow and Helge Kastrup

Chapter 13

Information technology: use, under-use, over-use, misuse

IT is a two-edged sword. It provides many tools to facilitate work in science. Looking for and handling information, doing computations, making measurements and presenting them have been revolutionized within a generation and have totally changed the landscape of physics teaching. So has the landscape of IT-based crime. And so have the psychological consequences for adolescents in 24–7 electronic contact with all their acquaintances.

13.1 Use

Here is an annotated series of IT appearances in the classroom, where each was a step forward. Many items in the list are so much a part of everyday life—they are on our smartphones—that one might forget what a change they represented[1].

Pocket calculators. Before we had these, we needed a slide rule, which could multiply, divide, and take square roots with an accuracy of two to three decimal points. We could not add or subtract. Logarithms, exponential functions, and spherical or hyperbolic trigonometric functions needed tables.

Programmable calculators (at the beginning only programmable in 49 steps). These were fit for writing lab reports, where one often had a column to fill in as a function of other columns of measurements, constants, and other quantities[2].

---

[1] The categories are not distinct. Most items on the list are included in the latecomer *smartphones/tablets*, and there are other overlaps.
[2] HK: I had in those days a clever student whose lab reports were suspiciously good, with nice standard deviations, slopes of graphs, and whatnot. Once I gave him an apparatus that was designed to give half the table value of Planck's constant h. His report found $h = (6.6 \pm 0.1) \times 10^{-34}$ J·s where the table value is $(6.626) \times 10^{-34}$ J·s. He admitted that he always doctored his measurements to give good values with reasonable spread with the help of a pocket calculator program that he had written.

Desktop computers with text editing, spreadsheets, graphic programs, statistic programs, PowerPoint presentations, and stronger math programs such as Mathcad, Mathematica, Maple, GeoGebra, and Knime, just to mention a few. In addition, they provided access to huge data sets. Students could make mathematical calculations that were normally too difficult or too time-consuming—to the point of impossibility—to solve with pencil and paper.

The Internet, with access to a smorgasbord of information, but with no guaranteed peer review. This is in itself a subject for science teaching.

Social networks such as Twitter, Facebook, Messenger, Instagram, and TikTok are present in class in the students' minds, and should be taken seriously. They can be used to form a unifying identity for a class, a team, or a group: *us* in contrast to *them*.

E-mailing and *texting* can speed up communicating between and among students.

Electronic conferencing through Facetime, Skype, WhatsApp, Messenger, and many others. This is good for groups working outside of lab, and for group work generally, not the least in times of COVID-19 or other diseases.

Applets and other simulation programs. These are quite important for science-anxious students who fear electronic equipment, and for those getting anxious in the tedious step between equations and models. Today many of these work on smartphones/tablets.

YouTube has (in addition to music videos, puppies falling in the soup, cats meowing at a vacuum cleaner) a plethora of tubes explaining most scientific subjects as lectures, animations, films of nature phenomena, and tubes explaining everything from how to change your gas bottles to how to grow your own crystals.

Smartphones/tablets that, in addition to many of the features already mentioned, include photo and video cameras. As for degrees of freedom, (see chapter 9), pictures and videos will encourage more options for how students choose to present their lab work.

Dataloggers are available today for many scientific purposes; for example, pH-measurements, chemical concentrations (oxygen, carbon dioxide, salinity, etc), temperature, conductivity, light intensity, speed, acceleration, distance, pressure, voltage, and current. One can choose to make long series of measurements. Dataloggers can register measurements directly in spreadsheets, graphic programs, and math programs. Students can make measurements on short and long timescales. Here are a few examples:

1. If light intensity is measured every 0.1 ms, you can see a light bulb's output vary with the frequency of the AC source. It's a bit of an Aha!-moment watching it. Figure 13.1 shows such a measurement (from Denmark, where the AC frequency is 50.0 Hz).
2. With an oxygen probe in an aquarium with water plants, you can follow the oxygen content increasing during the day and decreasing during the night.
3. With a GM-probe you can measure the background radiation level over a whole week.

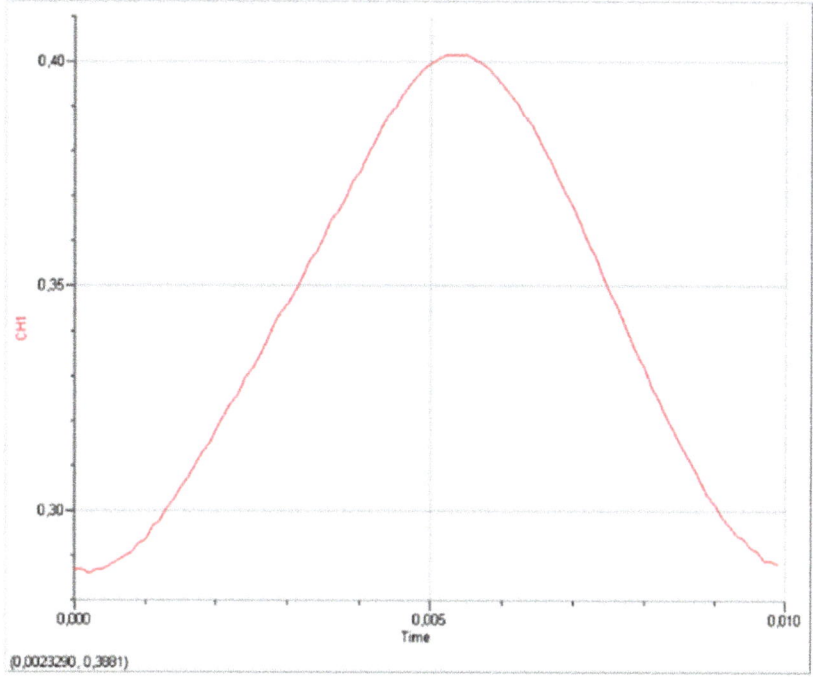

Figure 13.1. Light bulb output versus AC frequency.

IT technology can be both the problem and part of the answer to anxiety. If, for example, dataloggers are treated like other complicated equipment, this can easily provoke the 'Verfremdung' (alienation) causing science anxiety. If students are scared of assembling an electric circuit with analog meters, there is no indication of their being less scared with digital dataloggers. So all the precautions taken against science anxiety in traditional lab work (see chapters 9 and 10) must be taken advantage of.

On the other hand, many IT-based methods or appliances have the potential to ease students' fear and anxiety. Here are some examples:

- Instead of wondering how to connect equipment, watch a YouTube video.
- Don't ponder how to visualize curved space in the general theory of relativity, watch a YouTube video.
- To show a presentation about how light is absorbed by chlorophyll, make a film with a digital microscope and show how the chlorophyll is moved around in the leaves. It is another Aha!-moment the first time they watch it: 'Science is fun!'
- It is a major change in lab work to include how results are presented as a degree of freedom (see chapter 9). There is absolutely nothing wrong with traditional written reports. Those pocket calculators, spreadsheets, graphic programs, and statistical programs are of great help in doing so, simplifying computations and graphics. But apps such as TikTok, Instagram, or just the

smartphone's camera will help to make presentations more interesting. Plus, they are probably better known to the students than to the teachers. They are an important part of students' daily life. Presentation programs such as PowerPoint are also ready to use straightaway when you learn how (which is not rocket science!). But students are more familiar with apps known from their smartphones.

- If students can find a datalogger app to smartphones, they can use it whenever possible. They can also get dataloggers whose output is directly connected to a smartphone. Today, all students use their internal stop-clock in their smartphones instead of traditional analog stop-clocks. Here are a few existing physical measurements they can make from various apps: time, position, and speed[3]; frequency, amplitude, and level of sound; light intensity; magnetic field strength. In addition to the students' familiarity with smartphones (and therefore less danger of their getting anxious), a further advantage is that they can go directly out into their everyday life and measure various physical quantities (thus learning that physics is not an abstract theoretical discipline).
- The use of virtual reality has become a standard component of physics teaching, both for secondary schools and universities [1]. But with it comes anxiety. Here is a description by a sales consultant at a leading global technology company. He presents virtual reality experiences to business professionals:

  'I interact with professionals across industries. Virtual reality is a large part of my job. I had a group of business executives to whom I gave a demonstration. I asked them what their experience was with virtual reality. One of them told me that their company was experimenting with having meetings with virtual reality rather than Zoom. The problem that came up were that most of the people didn't know how to use it, which immediately caused anxiety for people who were just trying to participate in a meeting. They spent most of the meeting trying to figure out how to use it, rather than participating. As you would guess, they ended up feeling that other people in the meeting knew what they are doing better than they themselves did. This led them to feel anxious and behind the times with technology'. [2]

If this is the case with seasoned professionals, it will be demonstrably the case with students, for some even worse. Video gamers will have a leg up from the start. Teaching will then have to include extra sessions for those students who do not play video games. These sessions will also have to address the concomitant anxieties.

---

[3] From their GPS.

13.2 Under-use

If you don't use the items on the list above where appropriate, it is 'under-use'. The problem is the word *appropriate*. One can always argue that looking up facts in Britannica instead of using the Internet is choosing a peer reviewed source. (Well, Britannica is also to be found on the Internet, and few labs have a paper edition of Britannica from this century.) But not using up-to-date technology with which the students are familiar is under-use.

13.3 Over-use

Even though we have stressed the importance of using smartphones/tablets because of the students' familiarity with them, there are important considerations to be made regarding their presence in the classroom. Neuroscientists have used the word *addiction* describing young persons' need to always be in contact with their peers by continually checking their phone[4]. The phones typically transmit sound signals when receiving messages from e-mail, WhatsApp, Instagram, TikTok, and Messenger. When muted in class, they are often on vibrate mode in the pocket, so the owners are still aware of what arrives.

Thus, the over-using pattern of smartphones involves a tendency to check notifications all the time. The human brain is never really able to multitask. So consciousness must be divided between the phone and the class. Research also substantiates that over-use of smartphones can lead to anxiety (our major theme), depression, and loneliness. It also shortens the attention span and makes it harder to concentrate for longer periods.

Because of these problems, phones have recently been forbidden both in France up to class 9 (nine-year-olds) and in primary schools in the Australian state Victoria. In both places the decision was controversial but was generally supported by a majority of parents.

13.4 Misuse

We will not cover misuse such as digital counterfeiting, theft of hardware and software, hacking, phishing, illegal copying of music or films, stalking, or use of obscene or abusive language.

Legal misuse. At all times we know of examples of theses, papers, assignments, and homework with doctored experimental results, illegal long quotes, and other sorts of cheating. Such fraud is easier today with cut-and-paste techniques. And the hunting ground for quotable stuff is ever growing and increasingly accessible through the Internet. Fortunately, all teaching institutions have software for caching the culprits. But some written materials still slip through the sieve. We must teach how and why offenders are or should be treated. As good quality artificial intelligence is now available free of charge, students could be tempted to use it to write their homework. We are convinced that this problem very soon will pop up as

---

[4] *Nomophobia* is the clinical word for the fear of being without your mobile phone. (Really!)

a major issue. Should it be forbidden? How do you do that? Is it cheating? Is it causing anxiety for those that don't do it, for those that don't know how to do it, and for those who do it but can't understand what they write in their own homework?

Didactic misuse. There are two pitfalls to avoid when teaching students to use a computer program or other electronic device:

1. The Pinocchio error (also called the marionette puppet error). You tell the whole class to work in unison following your instructions. 'Push button A—move the mouse to the red box and click the right mouse button—blah blah blah.' It is a boring approach, and it won't work. Within half a minute, the IT-knowledgeable students will be far ahead, and have started checking their e-mails, while others are already lost in the digital woods.
2. The cookbook error. You have written down the instructions using the Pinocchio method. It won't work. In tradition-speak, 'Boys[5] don't follow an instruction book'—if it isn't self-instructive, it isn't worth following. Speaking in more up-to-date language: Most IT-functions are so standardized that students of the 2020s know them already. F1 is 'help', clicking the left mouse button is for choosing the next option, and so forth.

What then should the instructor do? Assign them what tasks need to be done, what to find out, what to measure. Show them available programs, utilities, and apparatus. And remind them that they can talk together and that you the instructor are present in the room. The goal is to teach and to model appropriate use of IT, providing both knowledge and confidence.

References

[1] Search results: virtual reality in physics education *Google Scholar* https://scholar.google.com/scholar?q=virtual+reality+in+physics+education&hl=en&as_sdt=0&as_vis=1&oi=scholart

[2] Mallow D 2022 private communication

---

[5] And not only boys.

Chapter 14

Online education: interaction at a distance

Online learning even under normal conditions has produced new challenges to physics pedagogy. To give just a few examples, teacher–student interactions become less personal, students' engagement in group work can diminish, and communication by use of whiteboards, tablets, and other devices can be daunting. Studies of distance learning and its effect on students during the pandemic have been undertaken. We shall analyze how our earlier studies of attitudes and anxieties can be reproduced for distance learning.

Most publishing houses selling textbooks will have homepages associated with the books. Here you will find extra material, such as, *inter alia*, answers to selected questions, further exercises, and video material. Likewise, national associations, ministries of education, and other organizations such as space agencies have homepages with valuable educational material. Although highly important, these homepages are not our focus. They are ancillary tools. We will concentrate on the direct communication between teacher and student online.

In Denmark teaching from primary schools up to university level has functioned online in 2020 and part of 2021 during the COVID-19 pandemic. Many different platforms were used. We will not discuss the merits and demerits of those, as the market changes quickly, so our reviews will soon be obsolete.

The amount of online education is increasing, and not only for health reasons. It is a way to reach students living in less populated regions and thereby minimizing travel and accommodation expenses. It is important for in-service training, where students can study at their own pace and hours. Globally, many distant places with no universities and no money for traveling are already being helped with online education, and many more will in the years to come. And today with access to local solar and wind power and much more easily distributed Internet, earlier obstacles are diminishing. This is important in rich countries such as Denmark, the US, and the UK, as well as in poorer countries. And there is no doubt that the art of online teaching is in for a huge development. Already today it is an industry of its own.

Many Western universities make money selling courses they are already doing anyway. Many platforms offer online teaching on a plethora of subjects and levels where you pay by the hour. Examples of very popular subjects are English language courses for Chinese students, as well as science courses, and math courses in calculus and algebra. The prices vary (who would have guessed?) according to the teacher's credentials and student evaluations.

14.1 Methods of online education

Generally speaking, we have five different methods of online education:

1. Videotaped teaching (lectures, experiments, etc) to be watched by students when they choose.
2. Teaching streamed with a single or very few students in interaction with the instructor.
3. Teaching a distant class with interactive cameras and microphones both ways.
4. Teaching a class where students are in their own homes, or equivalently elsewhere, with interactive cameras and microphones both ways. The students might or might not be able to communicate with each other.
5. Webinars—web seminars, are the online version of one form of IE: a mixed lecture and interactive format.

14.1.1 Videotaped teaching

During the COVID-19 pandemic my (HK) classes at the university of Copenhagen were divided in two groups having their lectures alternate weeks to reduce infections. The lectures were recorded including questions from the students, and their answers. Therefore, everybody could follow the full program. So could those ill or absent for other causes. This service will go on after the pandemic is over, for obvious reasons. One might argue that it would enable lazy students to play truant and thereafter not follow the lectures they missed. But we regard it as more important to help the diligent students.

Educational systems worldwide are pressed economically. It might be tempting for administrations to save money by using the same filmed lectures again and again for years. We can all treasure especially gifted lecturers like Richard P Feynman performing on videos. But Feynmans are rare. For most students, being present, catching the eyes of the lecturer, laughing with the audience at her jokes, asking questions, and being challenged by her is the best pedagogy, notwithstanding the merits already mentioned. It is not a good idea to replace normal teaching with videos. And experimental work must be done in the lab.

YouTube has a great number of teaching videos on nearly all subjects relevant for readers of this book: introductions to general relativity or the Dirac equation, understanding what black energy and black matter are at one end of the scale, to much more mundane questions like Plimsoll lines, annihilation and pair production, or Stokes' law for laminar flow. Such videos are important for students' private study and are often recommended by lecturers. The same goes for videos explaining the functions of simple or more complicated apparatus such as microscopes, voltmeters, stroboscopes, etc. If the class is working with open experimental questions, they can be very useful, as can the so-called applets, presenting thousands normally very simple simulations of phenomena, e.g. ripple tank analogues, dispersion, gas molecules, projectile motion, Kepler's laws, diffraction by single or double slits, and the photoelectric effect.

Generally speaking, videotaped lectures are best used alone or in small groups working together, so students can stop to take notes, go back for missed information and fast forward at well-known material. (Another warning against letting taped lectures replace live lecturers!) And we recommend lectures being made available for students who were absent or who want to review.

14.1.2 Teaching a few students with interaction

It is important to know that students looking for online help often are anxious of science beforehand. A gifted teacher can help anxious students over their fear with personal charisma and empathy; the online teacher can do the same. But it requires a stronger presence, and you have to be extra well prepared so you can improvise according to the students' reactions that are obviously harder to monitor.

14.1.3 Teaching a distant class

HK taught for some years from Copenhagen a class of students living on the island of Bornholm, which he has visited either with airplane or car, or ferryboat plus car. For experimental work he had to use a car to bring equipment and stay for the night. Flying was quite expensive. Therefore, normal classroom teaching was often done digitally. Both at his institutional home, University College Copenhagen, and at the institution on Bornholm there were some fancy (and expensive) cameras. The students could zoom, refocus, and point the camera and adjust the sound level, and so could he the other way around. Obviously, this requires a high-end Internet connection. He knew the class well and they knew him. Most of the time he lectured, with students doing group work, with Q&A sessions, and a few experimental demonstrations. So, with the prerequisites of knowing the students beforehand, doing a limited amount of experimental work, and having the right equipment, this is possible and works well. We even used the system for a number of oral Bachelor's degree examinations without problems. This requires local control to avoid illegitimate help from fellow students or others. Both the external examiner and HK were in Copenhagen, and the students on Bornholm.

14.1.4 Teaching students, each sitting at home

Professional programs galore offer easy ways to make tests of the type 'Here are 20 countries and 20 capitals. Move the cities with your mouse so that France and Paris pair, Italy and Rome pair' and so forth. The programs give immediate response—the number of correct answers. So the instructor can work with hundreds of students with only little effort. But the problem is that no real learning is occurring. And the moderator is not teaching. Technically there is some mnemonic help in learning the capitals of European countries, but why use online education for that? And why pay for it? To us, this kind of interaction is fake.

Talking to a group of more than a few students on the screens sitting in their homes is extremely demanding and exhausting for a teacher. Lecturing without interactions works, but with limits. It can succeed if the group members know each other well. If not, it is hard work to establish a group consciousness, a 'we' feeling. It easily becomes impersonal. And the teacher easily loses the overview of whether all participants are active or only sleeping partners.

14.1.5 Webinars

The online version of a simple IE—lecture plus class discussion—the webinar has limitations. Some are elucidated in the previous paragraph. Others are the constraints on some of the methods we discussed in earlier chapters. Perhaps the most obvious is the elimination of our modified Socratic method (see chapter 11). It is simply not possible to keep track online of who raises their hand most—or more important, least—nor for the other students to watch the teacher–student interaction at the same time. At best it can imitate simple IE. While perhaps unavoidable due to technology limitations, it should be used only when necessary.

14.2 The online environment

The environment in which online learning occurs must be both comfortable and safe. The former is an issue that has been, and is still being dealt with. The latter is the new feature due to pandemic conditions. In the beginning of the chapter, we discussed the former. Here we focus on the latter. We are assuming that the current plague is not going to be a one-off. Monkey pox is already upon us, and who knows what's next? Thus, we must be prepared to modify physical environments; students must at minimum be safe and feel safe. After that we can deal with fear of physics.

4.2.1 Bricks and mortar

- Buildings and the rooms therein can no longer be built to house the same number of students as before. Their number density must decrease.
- Several students at a time cannot sit around one computer. Social distancing must be built in.
- Local courses where an online local class requires one or more in-person meetings will need to be eliminated.
- Internet cafes must have well separated carrels with plastic shields around each.
- More interactive computers must be provided to people who cannot afford them.

It is obvious that this will take considerable funding, which must be provided by governments and foundations, with the richer helping the less fortunate.

4.2.2 Social interaction

As all good teachers know, eye contact is the *sine qua non* of success. And it is severely compromised by masks. (That's why robbers wear them. Whoever thought we'd see the day when we could walk into a bank in a mask, and ask for money?) How long this will endure is anyone's guess. We will have to live with it intermittently for the foreseeable future. But many online interactions, as described in section 14.1, have already eliminated that feature. Group projects, in and outside class, also discussed above, will have severe limitations. There are more.

And yet, with the curse comes a blessing. We can interact with others around the world from the safety of our own home, or wherever we have put our computer. International conferences don't have to 'be' anywhere. Researchers of course have been interacting online since it became practical. Guest speakers are much cheaper *sans* travel and accommodations.

4.3 What can (and cannot) be done?

As we and other physicists have been railing against as long as we can remember—no more huge lecture halls[1]. Hake found (chapter 11) that attempts to utilize IE were overwhelmed by sheer class size. Happily, social distancing rules will put an end to those, as long as the institution is consistent in their application. That alone reduces a major source of anxiety in science classes.

Group projects present a particular hurdle; first, because professional physics itself is mostly done in groups; second, because essential hands-on laboratory interaction is not possible; last but not least, females are more confident in group rather than traditional learning (chapter 6). Here we can only hope that a decrease in frequency and severity of pandemics will make these again feasible. We can of course just go with social distancing in lieu of online learning and hope for the best, but the game may not be worth the candle. Perhaps simulations can provide something of a substitute. They have certainly proved effective in state-of-the-art science and engineering applications, such as modeling the activity of the US electrical power grid [1].

Then there is the issue of cheating. Online education has had this from the start. Students collaborate with each other on homework and exams. Who actually solved the exam, the student or his older sister with a PhD? Perhaps worst are the companies that quickly acquire copies of tests and results and post them ostensible to aid student learning. The pandemic has made things worse, as more and more education shifts online. The August 2022 edition of *Physics Today* [2] describes the various methods of online cheating, and how teachers can try to overcome them. But the problems are far from solved. One situation screams for rectification. We mentioned it before. We quote from the article: '[I]n introductory classes at large institutions, classes can have hundreds of students, and tests are typically multiple choice.' Whatever that is, it's not physics education. It should be abolished. Don't do it! Institutions need to have the means to lower class sizes (pay the coaches less at US universities?).

---

[1] And that goes for you too, chemistry and biology!

Large lecture classes are not just prone to cheating, they also undercut IE, the proven method for educating physics students beyond what a lecture can do. Whether the lecture is online or offline is moot. The anxiety production is the same. Thus, the methods we have developed so far for combatting it can be the same. But additional confidence-building methods will have to be designed. No matter what the format, the lack of physical presence, with its immediate unmasked face-to-face interactions of the whole class, cannot generally not be reproduced, not even on Zoom or its equivalents.

References

[1] Chassin D P, Schneider K and Gerkensmeyer C 2008 GridLAB-D: an open-source power systems modeling and simulation environment *2008 IEEE/PES Transmission and Distribution Conf. and Exposition* pp 1–5
[2] Feder T 2022 College instructors adapt their teaching to prevent cheating *Phys. Today* **75** 25–7

IOP Publishing

Fear of Physics
And how to help students overcome it
Jeffry V Mallow and Helge Kastrup

Summary

In this book, we have described the nature of the fear of physics, its descriptive causes, and its prescriptive interventions. We have quoted our research and that of others, as well as classroom experiences. We have seen that there are multiple issues at play. In chapters 2, 3, and 5, we have enumerated the connection between attitudes and anxieties. We found that one of the predictors was gender, and have devoted chapter 6 to this. We have also found that another predictor was non-science anxiety, namely anxiety about other subjects. The reason for this is not clear, although 'trait anxiety' probably lies behind 'state anxiety'[1]. However, the third predictor was choice of field of study, which prompts us to ask, have students chosen the non-sciences purely out of interest, or because of anxieties and/or poor physics pedagogy?

In chapter 4 we discussed the Janus-like nature of constructivism. On the one hand, physics pedagogy needs to be, and to a large extent has become, constructivist. Straight lecturing has been shown to yield poorer learning outcomes than interactive engagement. But the movement called radical constructivism has an altogether different agenda: to undercut science in general and physics in particular, by attacking the existence of objectivity. Unfortunately, the belief that 'there are no facts' has trickled down to our students, shaping their attitudes, which correlate with anxiety.

An additional feature of radical constructivism is its claim that physics is intrinsically anti-female and women should avoid studying it. In chapter 6 we showed that this is demonstrably false. We compared females' study choices and discussed their enrollments, focusing in particular on why percentages have stagnated. We outlined proposed remedies from various countries and described some that have shown success.

In chapter 7 we examined whether nationality affects science attitudes and anxieties. We first examined the Programme for International Student Assessment (PISA) results. We then focused on our studies of attitudes and anxieties for our two

---

[1] Trait anxiety is an intrinsic predisposition. State anxiety is provoked by, for example, poor physics teaching.

countries. Both showed that the primary factor was the quality of teaching. We also found from our studies of American and Danish students that nationality plays some role, due to differences in the educational systems and in the societies as a whole. Nevertheless, there are some commonalities, especially with regard to anxiety.

In chapter 8 we compared math anxiety to science anxiety. While these are distinct phenomena, and should be treated separately, math anxiety predisposes secondary school students to avoid both math and physics. This carries over to the university, constricting the choices of study and the opportunities for careers involving mathematics and/or physics. Math anxiety and science anxiety clinics have proven to be successful interventions. Teachers' attention to the issue is also essential.

Chapter 9 dealt with laboratory pedagogy. We discussed the structure of experiments, focusing specifically on 'numbers of freedom' analogous to 'degrees of freedom' in mathematics and physics; namely, the distance from cook-booking. We also considered the structure of the groups carrying them out. Issues of gender equity were addressed. We described the effect of all of these on student attitudes and emotions, and prescribed modifications in pedagogy.

In chapter 10 we discussed group project work at various institutions in both Denmark and the United States. These included a gymnasium, a teacher training college, and two universities. In addition to listing and explaining numerous projects, we focused on confidence-building for both females and males.

In chapter 11 we considered techniques for teaching different groups of students: physics majors, pre-health students, and liberal arts/social science students. In all cases, interactive engagement (IE) gave better results than strictly traditional lecturing, as a course component or on its own. Naturally, the course topics and the mathematical requirements were not identical; nevertheless, the pedagogy can and should be similar. In the case of physics students, contrary to popular opinion, IE did not reduce coverage. We described various forms that IE can take, tailored to each student population.

In chapter 12 we addressed the connections between physics and society. We considered the need for physicists to make their field 'user-friendly', to attract all kinds of students. We described the spectrum of nerds through anti-science pseudo-intellectuals. Of the last, we despair, short of an epiphany on their part. Finally, we described what physics can do *for* society, providing correct information, interacting directly with communities by connecting course topics with issues of critical importance, engaging in research addressing societal issues, and involving students in studying and carrying out projects on specific topics, such as energy.

In chapter 13 we listed and described the many uses of information technology. We considered over-use, under-use, and misuse. We enumerated the possible sources of anxiety and prescribed methods for amelioration.

In chapter 14 we dealt with online education. We examined the forms it can take in physics education, its advantages, and its pitfalls. We demonstrated the ways in which it cannot replace face-to-face learning in the classroom, in the laboratory, and in group project work. We also considered ways in which current online methodologies can partially replace them. Throughout, we focused on building student confidence in the face of new obstacles.

We know of course that the issues we have raised, the explanations we have provided, and the prescriptions we have proposed may be ignored or even opposed. However, we hope that the heads of programs, departments, divisions, and even the leadership of the university, will use the book and act on it. Many inroads into combatting math anxiety and general science anxiety have been made. The template is there.

At the outset of the book we asked, why physics? Well, why not physics?

IOP Publishing

Fear of Physics
And how to help students overcome it
Jeffry V Mallow and Helge Kastrup

Appendix

Questionnaires

A.1 Student attitudes questionnaire ([1] with permission of Springer Nature)

Instructions. Please circle the number that best describes the degree with which you agree or disagree with each item below, using the following scale:

Strongly disagree	Disagree	Neutral	Agree	Strongly agree
1	2	3	4	5

1. Science reflects the social and political values, philosophical assumptions, and intellectual norms of the culture in which it is practiced.
2. Science is a 'level playing-field' in which men and women have equal status and opportunity.
3. Tomorrow's truths in science will be different from those of today.
4. It is possible for two scientists to carefully perform the same experiment and get very different results, each of which is correct.
5. Science has nothing to do with my life.
6. Scientists agree on fundamental subjects like global warming, disposal of nuclear waste, and the use of stem cells.
7. Science is by its nature hostile to women.
8. Newton's laws of motion may eventually be modified.
9. Scientists' ideas apply to some physical objects in the Universe but not others.
10. The difference in number of men and women scientists is primarily due to biological differences.
11. The choice of topics for scientific research is affected by the views of the culture in which scientists work.
12. There are no such things as objective facts.

13. Science is boring.
14. The difference in number of men and women scientists is primarily due to differences in opportunity.
15. Science is inherently cold and unfriendly.
16. Science is a conspiracy between governments and scientific agencies formed to keep ordinary people from taking part in the democratic process.
17. Although interpretations can be ambiguous in things like personal relationships or poetry, in science the facts speak for themselves.
18. Newer scientific theories build on their predecessors.
19. Scientific experiments do not really discover 'the laws of nature', but instead let scientists find whatever they expect or want to find.
20. Women have a harder time succeeding in science than men do.
21. Modern scientists are responsible for most of the dangers in our world.
22. Science is a mental representation constructed by the individual.
23. When it comes to controversial topics such as which foods cause cancer, there's no way for scientists to evaluate which scientific studies are the best: everything's just a matter of opinion.
24. Every scientific theory is eventually proved completely wrong, and must be discarded.
25. The scientific view of the world is just an agreement among scientists.
26. Despite what scientists would have us believe, science is actually subjective.
27. Science transcends national, political, and cultural boundaries.
28. Scientists don't understand normal people.
29. The natural world can best be explained through a combination of perspectives, including science, paranormal phenomena, and astrological horoscopes.
30. The difference in number of men and women scientists is primarily due to individual choice.
31. The scientific knowledge in use today may be obsolete tomorrow.
32. Scientific laws work well in some situations but not in others.
33. Current ideas about particles that make up the atom will always be maintained as they are.
34. Objective facts are an illusion.
35. I cannot fulfill my need for creativity within the closed framework of the natural sciences.
36. Science is a naturally male field.
37. Scientific theories are simply agreements among scientists.
38. Current ideas about particles that make up the atom apply to physical objects everywhere in the Universe.
39. The reason fewer females than males choose careers in science is that women don't want to be restricted to the narrow scientific way of understanding the world.
40. The results of scientific research experiments are affected by the views of the culture in which scientists work.

A.2 Science anxiety questionnaire (Reproduced with permission from [2])

The items in the questionnaire refer to things and experiences that may cause fear or apprehension. After each item, place a number that describes how much YOU ARE FRIGHTENED BY IT NOWADAYS.

0	1	2	3	4
Not at all	A little	A fair amount	Much	Very much

1. Learning how to convert Celsius to Fahrenheit degrees as you travel in Canada.
2. In a philosophy discussion group, reading a chapter on the categorical imperative and being asked to answer questions.
3. Asking a question in a science class.
4. Converting kilometers to miles.
5. Studying for a midterm exam in chemistry, physics, or biology.
6. Planning a well-balanced diet.
7. Converting American dollars to English pounds as you travel in the British Isles.
8. Cooling down a hot tub of water to an appropriate temperature for a bath.
9. Planning the electrical circuit or pathway for a simple 'light bulb' experiment.
10. Replacing a bulb on a movie projector.
11. Focusing the lens on your camera.
12. Changing the eyepiece on a microscope.
13. Using a thermometer in order to record the boiling point of a heating solution.
14. You want to vote on an upcoming referendum on student activities fees, and you are reading about it so that you might make an informed choice.
15. Having a fellow student watch you perform an experiment in the lab.
16. Visiting the Museum of Science and Industry and being asked to explain atomic energy to a 12-year-old.
17. Studying for a final exam in English, history, or philosophy.
18. Mixing the proper amount of baking soda and water to put on a bee sting.
19. Igniting a Coleman stove in preparation for cooking outdoors.
20. Tuning your guitar to a piano or some other musical instrument.
21. Filling your bicycle tires with the right amount of air.
22. Memorizing a chart of historical dates.
23. In a physics discussion group, reading a chapter on quantum systems and being asked to answer some questions.
24. Having a fellow student listen to you read in a foreign language.
25. Reading signs on buildings in a foreign country.
26. Memorizing the names of elements in the periodic table.

27. Having your music teacher listen to you as you play an instrument.
28. Reading the theater page of *Time* magazine and having one of your friends ask your opinion on what you have read.
29. Adding minute quantities of acid to a base solution in order to neutralize it.
30. Precisely inflating a balloon to be used as apparatus in a physics experiment.
31. Lighting a Bunsen burner in the preparation of an experiment.
32. A vote is coming up on the issue of nuclear power plants, and you are reading background material in order to decide how to vote.
33. Using a tuning fork in an acoustical experiment.
34. Mixing boiling water and ice to get water at 70 °F.
35. Studying for a midterm in a history course.
36. Having your professor watch you perform an experiment in the lab.
37. Having a teaching assistant watch you perform an experiment in the lab.
38. Focusing a microscope.
39. Using a meat thermometer for the first time, and checking the temperature periodically till the meat reaches the desired 'doneness'.
40. Having a teaching assistant watch you draw in art class.
41. Reading the science page of *Time* magazine and having one of your friends ask your opinion on what you have read.
42. Studying for a final exam in chemistry, physics, or biology.
43. Being asked to explain the artistic quality of pop art to a 7th grader on a visit to the art museum.
44. Asking a question in an English literature class.

A.3 Student interview questions ([3] with permission of Springer Nature)

1. What choices for advanced study will you make, and why?
2. How many of your teachers in science have been male? Female? How has that affected you? Do they teach differently?
3. Do you find any of the sciences dry? If so, which ones and why?
4. Has your mathematics experience had an effect on your attitudes towards science?
5. What causes of anxiety in science can you identify?
6. Why do you think gender differences exist in subject choices?
7. How do your non-science friends view you? OR How do you view science students?
8. How are you affected by others' views of scientists and science students? Is this gender related?
9. What types of gender interactions have you experienced in science classes?
10. What do you think of single-sex educational groupings?
11. Do you experience anxiety in any subjects or situations?
12. Some people have suggested that there are no such things as objective facts and that science is simply constructed from the personal opinions and

subjective beliefs of scientists. How do you feel about this particular viewpoint?
13. Some people have suggested that science is inherently hostile and biased against women. How do you feel about this particular viewpoint?
14. Some people have suggested that science is inherently cold, unfriendly, and negative toward the individual. How do you feel about this particular viewpoint?
15. Some people argue that science is one way of describing the natural world among others to be considered as well; for instance, astrology, creationism/intelligent design. What is your opinion of this?
16. Some people argue that all of the science of today will become obsolete and forgotten in centuries to come. What is your opinion of this?

A.4 Administrator interview questions (Reproduced from [4]. Copyright Morgan & Claypool Publishers)

1. Students answered the following questions with highest frequency:

    - (2) How many of your teachers in science have been male? Female? How has that affected you? Do they teach differently?
    - (6) Why do you think gender differences exist in subject choices?
    - (7) How do your non-science friends view you? OR How do you view science students?
    - (13) Some people have suggested that science is inherently hostile and biased against women. How do you feel about this particular viewpoint?
    - (14) Some people have suggested that science is inherently cold, unfriendly, and negative toward the individual. How do you feel about this particular viewpoint?
    - (15) Some people argue that science is one way of describing the natural world among others to be considered as well; for instance, astrology or creationism/intelligent design.
    What is your opinion of this?

Why do you think that was the case?

2. Why do you think that was the case?
3. Do you think that there have been changes over recent years? Would students have chosen other 'favorite questions' in earlier years?
4. Which questions would you have chosen? Why do you think so? (They could respond as to which they would have chosen as students, or which they would have chosen at present.)
5. Do you have any other comments?

A.5 American Association of Physics: Teachers' practice questionnaire (Adapted with permission from [5])

The questionnaire is designed to help you assess your teaching style, and consider modifying it if you deem it necessary.

Circle the letter or letters of the alternative(s) that most nearly describe your behavior in each situation. Please circle at least one response for each situation.

1. The lab performance of your students seems to be getting worse. More of them are not completing the experiments during the lab period. You would:
    A. Emphasize the importance of laboratory work and the necessity for accomplishing the lab assignments.
    B. Use conversation to show your concern and to encourage the students.
    C. Discuss the function of experimental work in science and work with the students to improve the quality of their performance.
    D. Let the students be responsible for themselves.
2. Your students bombed the first hour-long exam. You have stressed the grading standards for the course. Classroom participation is beginning to improve. Now the second exam is coming up. You would:
    A. Engage in helpful friendly interactions and continue to make sure all students are aware of your standards.
    B. Be satisfied with the improvement and take no further action.
    C. Praise the class for its improved participation and let each student know you care about his/her work.
    D. Emphasize the importance of good study habits and the completion of each homework assignment.
3. At the beginning of class one student says, 'Problem #7 is driving me up a wall!' Several students agree. Normally you have left them alone as their overall homework performance has been good. You would:
    A. Ask them which science principles can be used and show how these principles are related to problem #7.
    B. Encourage those having trouble to try it again.
    C. Present a hint before going on with the class.
    D. Explain that many students have difficulty with problem #7 and if they wish to work on it some more, they should please see you during office hours.
4. You are teaching the course for the first time. The students are doing badly on the exams; perhaps the course is moving too fast. You have decided to make some changes. You would:
    A. Invite class discussion in developing the change. Do not direct.
    B. Slow down and see what happens.
    C. Allow class to formulate the basis for change in classroom activities.
    D. Incorporate group recommendations but direct change.
5. Students have not been doing their homework; furthermore, they seem unconcerned about it, in spite of the fact that you have continually

reminded them that doing the homework problems is an important part of learning science.
   A. You decide to do something about this situation. You would:
   B. Reduce the number of problems assigned as homework.
   C. Set aside class time for problem solving, and continue to assign the same amount of homework.
   D. Increase the percentage of the final grade based on homework scores.
   E. Have a class discussion on why they are not doing the homework.
6. Your lecture classes have increased their enrollment by 40%. This will greatly increase your office visit load. You want to be both efficient and effective as an advisor. You would:
   A. Inform the students that they may see you at any time.
   B. Require students to make an appointment and come with specific problems.
   C. Assign each student a specific time for a conference.
   D. Meet with the students in small groups at a specified time.
7. Students have been complaining that science is too abstract. Some of them have asked that you include some applications in the course. You would:
   A. Pick out applications for inclusion in your lectures.
   B. Ask them to come up with a list of applications of interest to them and from this list they select the three most relevant. You then include these topics in your lectures.
   C. Ask the class for some applications of interest to them. You pick out a few for inclusion in your lectures.
   D. Explain to them that science is based on abstract reasoning and that once they understand the basic principles, the applications will be easier to understand.
8. Many of the students in your large lecture section miss your Friday afternoon class. You would:
   A. Accept smaller Friday afternoon classes.
   B. Ask the students who do attend for suggestions on ways to increase class attendance and use those suggestions that are consistent with course goals to stimulate interest.
   C. Inform students you will include material on exams which is only covered during Friday classes.
   D. Contact individual students and encourage them to come to class.
9. You are teaching the lecture portion of a large lecture-recitation course. A student comes to your office and complains about the recitation instructor who the student thinks is requiring too much work. You know from past experience that students in this section do very well on the exams. You would:
   A. Tell the student to settle this with the recitation instructor.
   B. Explain the recitation instructor's role and your expectations and suggest that the student return to see you if he is still having trouble after a few weeks.

C. Tell the student to keep trying.
D. Change the student's enrollment to a section taught by another instructor.
10. Your students have done well in the first portion of your course but their performance on the second portion is poor and getting worse. You would:
    A. Ask the students how they feel about their poor performance.
    B. Encourage students to study harder and then move on to the next topic.
    C. Re-normalize the exam scores so that no one gets too low a grade.
    D. Discuss with students in class how you can help them to meet your standards.
11. You are taking over the second semester of a year-long course. The instructor who had the students in the first semester is highly regarded and respected by her students. You would:
    A. Announce that you expect the same level of performance in the course you have planned.
    B. Ask the students what they liked about the first semester of the course.
    C. Discuss the past class with students and describe differences in your approach.
    D. Try to do the same as the previous instructor.
12. During a laboratory class, you notice that one group of students is having trouble completing an experiment. You know them to be capable students. You would:
    A. Ask them what needs to be done in order to finish successfully and on time.
    B. Explain the science concepts and experimental apparatus again.
    C. Inform them you expect them to finish the assignment on time.
    D. Tell them you will accept their report using the data they have.

Scoring

Item	Relationship			Content	
1	C	B		C	A
2	A	C		A	D
3	A	D		A	C
4	D	A		D	B
5	B	D		B	C
6	D	C		D	B
7	C	B		C	A
8	B	D		B	C
9	B	D		B	C
10	D	A		D	B
11	C	B		C	A
12	A	D		B	A

A.6 Critical incidence inventory

1. What was your most engaged moment?
2. What was your most distanced moment?
3. What was the most affirming or helpful action?
4. What was the most puzzling action?
5. What surprised you the most?

References

[1] Bryant F, Kastrup H, Udo M, Hislop N, Shefner R and Mallow J 2013 Science anxiety, science attitudes, and constructivism: a binational study *J. Sci. Edu. Technol.* **22** 432–8
[2] Alvaro R 1980 The effectiveness of a science therapy program upon science-anxious undergraduates *PhD thesis* (Chicago, IL: Loyola University)
[3] Mallow J, Kastrup H, Bryant F, Hislop N, Schefner R and Udo M 2010 Science anxiety, science attitudes, and gender: interviews from a binational study *J. Sci. Edu. Technol.* **19** 356–69
[4] Kastrup H and Mallow J 2016 *Student Attitudes, Student Anxieties, and How to Address Them: A Handbook for Science Teachers* (San Rafael, CA/Bristol: Morgan & Claypool Publishers/IOP Publishing)
[5] Fuller R, Agruso S, Mallow J V, Nichols D, Sapp R, Strassenburg A and Allen G 1985 *Developing Student Confidence in Physics* (College Park, MD: AAPT)

www.ingramcontent.com/pod-product-compliance
Ingram Content Group UK Ltd.
Pitfield, Milton Keynes, MK11 3LW, UK
UKHW050244150426
5217IPUK00005B/123